ENERGY POLICIES, POLITICS AND PRICES

INTERNATIONAL POLICIES AND ISSUES OF NUCLEAR SAFETY AND MANAGEMENT

ENERGY POLICIES, POLITICS AND PRICES

Additional books in this series can be found on Nova's website under the Series tab.

Additional E-books in this series can be found on Nova's website under the E-book tab.

ENERGY POLICIES, POLITICS AND PRICES

INTERNATIONAL POLICIES AND ISSUES OF NUCLEAR SAFETY AND MANAGEMENT

MATTHEW D. LEELAND
EDITOR

Nova Science Publishers, Inc.
New York

For permission to use material from this book please contact us:
Telephone 631-231-7269; Fax 631-231-8175
Web Site: http://www.novapublishers.com

LIBRARY OF CONGRESS CATALOGING-IN-PUBLICATION DATA

International policies and issues of nuclear safety and management / editor, Matthew D. Leeland.
p. cm.
Includes index.
ISBN 978-1-61761-596-2 (hardcover)
1. Nuclear reactors--Safety measures--International cooperation. 2. Nuclear power plants--Safety measures--International cooperation. I. Leeland, Matthew D.
TK9152.I5244 2010
363.17'99--dc22

2010031749

Published by Nova Science Publishers, Inc. † New York

CONTENTS

PREFACE

Currently, 437 civilian nuclear power reactors are operating in 29 countries, and 56 more are under construction. After the Chernobyl accident, representatives of over 50 nations, including the United States, participated in the development of the Convention on Nuclear Safety, a treaty that seeks to promote the safety of civilian nuclear power reactors. The convention has been in force since 1996. This book explores international policies and issues concerning nuclear safety and management

Chapter 1- After several decades of widespread stagnation, nuclear power is attracting renewed interest. New license applications for 30 reactors have been announced in the United States, and another 160 are under construction or planned globally. In the United States, interest appears driven, in part, by tax credits, loan guarantees, and other incentives in the 2005 Energy Policy Act, as well as by potential greenhouse gas controls that may increase the cost of fossil fuels. Moreover, the U.S. Department of Energy is spending several hundred million dollars per year to develop the next generation of nuclear power technology.

Expanding global access to nuclear power, nevertheless, has the potential to lead to the spread of nuclear technology that could be used for nuclear weapons. Despite 30 years of effort to limit access to uranium enrichment, several undeterred states pursued clandestine nuclear programs, the A.Q. Khan black market network's sales to Iran and North Korea representing the most egregious examples. Concern over the spread of enrichment and reprocessing technologies, combined with a growing consensus that the world must seek alternatives to dwindling and pölluting fossil fuels, may be giving way to optimism that advanced nuclear technologies may offer proliferation resistance.

Proposals offering countries access to nuclear power and thus the fuel cycle have ranged from a formal commitment by these countries to forswear sensitive enrichment and reprocessing technology, to a de facto approach in which a state would not operate fuel cycle facilities but make no explicit commitment, to no restrictions at all. Countries joining the U.S.-led Global Nuclear Energy Partnership (GNEP) signed a statement of principles that represented a shift in U.S. policy by not requiring participants to forgo domestic fuel cycle programs. Whether developing states will find existing proposals attractive enough to forgo what they see as their "inalienable" right to develop nuclear technology for peaceful purposes

Chapter 2- Currently, 437 civilian nuclear power reactors are operating in 29 countries, and 56 more are under construction. After the Chernobyl accident, representatives of over 50 nations, including the United States, participated in the development of the Convention on Nuclear Safety, a treaty that seeks to promote the safety of civilian nuclear power reactors.

The Convention has been in force since 1996. GAO was asked to assess (1) parties' views on the benefits and limitations of the Convention, (2) efforts to improve implementation of the Convention, and (3) how International Atomic Energy Agency (IAEA) programs complement the Convention's safety goals. GAO surveyed the 64 parties to the Convention for which it was in force at the time of GAO's review and analyzed the responses of the 32 that completed it, analyzed relevant documents, and interviewed U.S. and foreign officials.

Chapter 3- The creation of high-activity, long-lived radioactive waste is an inevitable consequence of generating electricity in nuclear power plants. It also is an inevitable consequence of engaging in a set of activities associated with national defense, ranging from propelling nuclear submarines to producing the fissionable materials needed to construct nuclear weapons. Early in the nuclear era, the very-longterm management and the ultimate disposition of those wastes was not a high priority. By the mid-1970s, however, most nuclear-capable nations had begun to focus more intently on developing plans to ensure over the very long term that the wastes would not endanger public health and safety or do serious damage to the environment.

Especially within the last decade, those efforts have benefitted from increasingly fruitful international cooperation and coordination. The International Atomic Energy Agency, an autonomous organization with a working relationship with the United Nations, carries out technical assistance programs and provides regulatory guidance to its members. It also supports the implementation of the *Joint Convention on the Safety of Spent Fuel Management and on the Safety of Radioactive Waste Management,* a treaty signed by 55 entities, including all but five of the 31 nations that now operate nuclear power plants. Under the auspices of its Radioactive Waste Management Committee, the Nuclear Energy Agency, part of the Organization for Economic Cooperation and Development, has established a number of subsidiary bodies that sponsor multinational exchanges among 28 industrial democracies

In: International Policies and Issues of Nuclear Safety... ISBN: 978-1-61761-596-2
Editor: Matthew D. Leeland

Chapter 1

MANAGING THE NUCLEAR FUEL CYCLE: POLICY IMPLICATIONS OF EXPANDING GLOBAL ACCESS TO NUCLEAR POWER

Mary Beth Nikitin, Anthony Andrews and Mark Holt

SUMMARY

After several decades of widespread stagnation, nuclear power is attracting renewed interest. New license applications for 30 reactors have been announced in the United States, and another 160 are under construction or planned globally. In the United States, interest appears driven, in part, by tax credits, loan guarantees, and other incentives in the 2005 Energy Policy Act, as well as by potential greenhouse gas controls that may increase the cost of fossil fuels. Moreover, the U.S. Department of Energy is spending several hundred million dollars per year to develop the next generation of nuclear power technology.

Expanding global access to nuclear power, nevertheless, has the potential to lead to the spread of nuclear technology that could be used for nuclear weapons. Despite 30 years of effort to limit access to uranium enrichment, several undeterred states pursued clandestine nuclear programs, the A.Q. Khan black market network's sales to Iran and North Korea representing the most egregious examples. Concern over the spread of enrichment and reprocessing technologies, combined with a growing consensus that the world must seek alternatives to dwindling and polluting fossil fuels, may be giving way to optimism that advanced nuclear technologies may offer proliferation resistance.

Proposals offering countries access to nuclear power and thus the fuel cycle have ranged from a formal commitment by these countries to forswear sensitive enrichment and reprocessing technology, to a de facto approach in which a state would not operate fuel cycle facilities but make no explicit commitment, to no restrictions at all. Countries joining the U.S.-led Global Nuclear Energy Partnership (GNEP) signed a statement of principles that represented a shift in U.S. policy by not requiring participants to forgo domestic fuel cycle programs. Whether developing states will find existing proposals attractive enough to forgo what they see as their "inalienable" right to develop nuclear technology for peaceful purposes remains to be seen.

GNEP has continued as an international forum under the Obama Administration, but the Bush Administration's plans for constructing nuclear fuel reprocessing and recycling facilities in the United States have been halted. Instead, the Obama Administration is supporting fundamental research on a variety of potential waste management technologies. Other ideas for limiting the expansion of nuclear fuel cycle facilities include placing all enrichment and reprocessing facilities under multinational control, developing new nuclear technologies that would not produce weapons-usable fissile material, and developing a multinational waste management system. Various systems of international fuel supply guarantees, multilateral uranium enrichment centers and nuclear fuel reserves have also been proposed.

Congress will have a considerable role in at least four areas of oversight related to fuel cycle proposals. The first is providing funding and oversight of U.S. domestic programs related to expanding nuclear energy in the United States. The second area is policy direction and/or funding for international measures to assure supply. A third set of policy issues may arise in the context of implementing the international component of GNEP or related initiatives. A fourth area in which Congress plays a key role is in the approval of nuclear cooperation agreements. Significant interest in these issues is expected to continue in the second session of the 111th Congress.

INTRODUCTION

Renewed interest in expanding the role of nuclear power in meeting world energy demand has also led to increased concerns about limiting the spread of nuclear weapons-relevant technology. Most of this concern focuses on the nuclear fuel cycle, which includes facilities that can be used to make fuel for nuclear reactors or materials for weapons.

After languishing for several decades, nuclear power in the United States appears poised for new growth, with license applications announced for up to 30 new commercial reactors. Two new U.S. uranium enrichment plants are currently under development in anticipation of an increased demand for nuclear fuel. However, no U.S. facilities are currently planned for reprocessing spent nuclear fuel—the separation of uranium and plutonium to make new fuel. Other countries provide commercial reprocessing services and, with several notable exceptions, have kept their commercial and weapons fuel cycles separate.

To reduce the likelihood that nuclear fuel cycle facilities could be used for weapons programs, several proposals have been made in recent years to discourage additional countries from developing uranium enrichment and reprocessing capability. Because a major justification for such facilities is to ensure fuel supplies for a nation's nuclear power plants, proposals to discourage fuel cycle facilities have focused on various alternative ways to guarantee supplies of nuclear fuel. These ideas have ranged from guaranteed access to foreign fuel cycle facilities to the establishment of nuclear fuel stockpiles, or "banks," under international control.

Under the Bush Administration, the U.S. Department of Energy (DOE) considered nuclear power to be "the only proven technology that can provide abundant supplies of base-load electricity reliably and without air pollution or emissions of greenhouse gases."[1] The

National Energy Policy Development Group recommended in 2001 that President Bush "support the expansion of nuclear energy in the United States as a major component of our national energy policy." About the same time, DOE created the Generation IV International Forum to collaborate with 10 other states in investigating "innovative nuclear energy system concepts for meeting future energy challenges." Congress has since appropriated hundreds of millions of dollars to support several programs related to the development of new nuclear power plants in the United States, including the Advanced Fuel Cycle Initiative and Generation IV. In passing the Energy Policy Act of 2005, Congress created new federal incentives for nuclear power plant construction. In February 2006, the Secretary of Energy announced the Global Nuclear Energy Partnership (GNEP) as part of President Bush's Advanced Energy Initiative.

GNEP has continued as an international forum under the Obama Administration, but the Bush Administration's plans for constructing nuclear fuel reprocessing and recycling facilities in the United States have been halted. Instead, the Obama Administration is supporting fundamental research on a variety of potential waste management technologies.

Concerns about the nuclear fuel cycle have been increased by several high-profile cases of subversion of ostensibly commercial uranium enrichment and reprocessing technologies for military purposes. In 2003 and 2004, it became evident that Pakistani nuclear scientist A.Q. Khan had sold sensitive technology and equipment related to uranium enrichment—a process that can be used to make fuel for nuclear power and research reactors, or to make fissile material for nuclear weapons—to states such as Libya, Iran, and North Korea. Although Pakistan's leaders maintain they did not acquiesce in or abet Khan's activities, Pakistan remains outside the Nuclear Nonproliferation Treaty (NPT) and the Nuclear Suppliers Group (NSG). Iran has been a direct recipient of Pakistani enrichment technology.[2]

The Board of Governors of the International Atomic Energy Agency (IAEA) found in 2005 that Iran's breach of its safeguards obligations constituted noncompliance with its safeguards agreement, and referred the case to the United Nations Security Council in February 2006. Despite repeated calls by the U.N. Security Council for Iran to halt enrichment and reprocessing-related activities, and imposition of sanctions, Iran continues to develop its enrichment capability.[3] Iran insists on its inalienable right to develop peaceful uses of nuclear energy, pursuant to Article IV of the NPT. Interpretations of this right have varied over time.[4] IAEA Director General Mohamed El Baradei has not disputed this inalienable right and, by and large, neither have U.S. government officials. However, the case of Iran raises perhaps the most critical question in this decade for strengthening the nuclear nonproliferation regime: How can access to sensitive fuel cycle activities (which could be used to produce fissile material for weapons) be circumscribed without further alienating non-nuclear-weapon states in the NPT?

Leaders of the international nuclear nonproliferation regime have suggested ways of reining in the diffusion of such inherently dual-use technology, primarily through the creation of incentives not to enrich uranium or separate plutonium. The international community is in the process of evaluating those proposals and may decide upon a mix of approaches.

Most of the proposals are not new, but rather variations of those developed 30 or more years ago.[5] In the 1970s, efforts to limit or manage the spread of nuclear fuel cycle technologies for nonproliferation reasons foundered for technical and political reasons, but many states were nevertheless deterred from enrichment and reprocessing simply by the high technical and financial costs of developing sensitive nuclear technologies, as well as by a slump in the nuclear market. Several developments may now make efforts to limit access to the nuclear fuel cycle more feasible and timely: a growing concern about the spread of enrichment technology (specifically via the A.Q. Khan black market network, as well as Iran's efforts); a growing consensus that the world must seek alternatives to polluting fossil fuels; and optimism about new nuclear technologies that may offer more proliferation-resistant systems. Central to the debate is developing proposals attractive enough to compel states to forgo what they see as their inalienable right to develop nuclear technology for peaceful purposes.

At the same time, there is debate on how to improve the IAEA safeguards system and its means of detecting diversion of nuclear material to a weapons program in the face of a worldwide nuclear power expansion.

This chapter is intended to provide Members and congressional staff with the background needed to understand the current debate over proposed strategies to redesign the global nuclear fuel cycle. It begins with a look at the motivating factors underlying the resurgent interest in nuclear power, the nuclear power industry's current state of affairs, and the interdependence of the various stages of the nuclear fuel cycle. A number of proposals have been offered that are aimed at limiting direct participation in the global nuclear fuel industry by assuring access to nuclear fuel supplies:

Year	Agency	Proposal
2003	IAEA	Would establish internationally owned fuel cycle centers.
2004	United States	Would keep uranium enrichment and plutonium reprocessing in the hands of current technology holders, while providing fuel guarantees to those who abandon the option.
2005	IAEA	Explored a variety of options to address front end and back end problems and their attractiveness to different groups of states, and surveyed past proposals.
2005	Russian Federation	Would establish international fuel cycle centers.
2006	United States	U.S. Global Nuclear Energy Partnership originally proposed that certain recognized fuel cycle countries would ensure reliable supply to the rest of the world in return for commitments to renounce enrichment and reprocessing; also proposed solutions for recycling of spent fuel and storage issues.
2006	U.S., U.K., Russia, France, Germany, and Netherlands	Six Country Concept would establish reliable access to nuclear fuel.
2006	Nuclear Threat Initiative	Promised $50 million for an international nuclear fuel bank under IAEA supervision provided another $100 million donated within two years and IAEA organizes implementation.
2007	United States	Revised GNEP would promote an international nuclear fuel supply framework (without explicit renunciation of fuel technology) to reduce proliferation risk and a closed fuel cycle featuring recycling techniques that do not separate plutonium.

RENEWED INTEREST IN NUCLEAR POWER EXPANSION[6]

Commercializing nuclear power has proved far more challenging than first envisioned. World nuclear capacity had reached about 200 gigawatts during the 1980s, but as confidence in nuclear power safety declined after accidents at Three Mile Island and Chernobyl, the rate of further capacity additions fell more than 75% during the following decade.[7] Today, nuclear power plants have a total capacity of about 373 gigawatts—providing about 15% of the world's electricity generation. Though a significant amount, it is far less than that projected 50 years ago. High construction and operating costs, safety problems and accidents, and controversy over nuclear waste disposal slowed the worldwide growth of nuclear power.

With uranium once considered a scarce resource, reprocessing and fast breeder technology was promised as a means of extending the energy remaining in spent nuclear fuel. In the 1980s, as the economics of advanced nuclear technology became questionable with declining fossil fuel prices and increased uranium supplies, national programs to develop fast breeder reactors came nearly to a standstill. Moreover, the plutonium fuel produced by breeder reactors drew strong opposition over its potential use in nuclear weapons.

In the past few years, however, the original promises of nuclear power have attracted renewed interest around the world. What has changed?

Volatile prices for oil and natural gas are a fundamental factor in national energy policymaking. Average world prices for a barrel of oil rose from below $10 at the beginning of 1999 to above $130 in mid-2008. They then declined to around $50 in early 2009 and rose to around $75 by early 2010.[8] U.S. natural gas prices have followed a similar track, along with prices of U.S. exports and imports.[9] As a result, national governments are searching for alternative energy sources, often including nuclear power. However, only 20% of the world's electricity generation is fueled by natural gas and 7% by oil (the majority of the world's electricity is generated from coal),[10] so nuclear power's ability to directly substitute for oil and gas is limited, at least in the near term.

For nuclear power to have a significant impact on oil demand, long-term changes in energy-use patterns would have to take place, particularly in the transportation sector. One possibility is that nuclear power plants could be used to produce hydrogen, which could provide energy for fuel-cell vehicles. The U.S. Department of Energy has been developing processes that could produce hydrogen in a high-temperature reactor, an effort that has continued under the Obama Administration. Another possibility is the commercialization of all-electric or plug-in hybrid vehicles that could be recharged with nuclear-generated electricity. But even if such technologies were to be successfully developed, it would take many years for the new vehicles and, in the case of hydrogen, fuel delivery infrastructure to have a significant energy impact.

Government policies aside, volatile oil and gas prices are heightening interest in nuclear power by improving current projections of nuclear power's economic viability. In the United States, natural gas has been the overwhelming fuel of choice for new electrical generation capacity since the early 1990s, although coal-fired plants still generate nearly half of U.S. electricity (and 40% of world electricity[11]). Because fuel costs constitute a relatively small percentage of nuclear power costs, higher natural gas prices could make new nuclear power plants economically competitive, despite higher uranium prices.[12]

Growing worldwide concern about greenhouse gas emissions, particularly carbon dioxide from fossil fuels, has renewed attention to nuclear power's lack of direct CO_2 emissions. President Obama's 2010 State of the Union address explicitly included "a new generation of safe, clean nuclear plants" as part of the Administration's energy and environmental strategy. Policies to reduce greenhouse gas (GHG) emissionsmay also indirectly encourage nuclear power expansion by increasing the cost of electricity from new fossil-fuel-fired power plants above that of nuclear power plants.

Some support for using nuclear power to reduce GHG emissions has emerged in academic and think-tank circles. As stated by the Massachusetts Institute of Technology in its major study *The Future of Nuclear Power*: "Our position is that the prospect of global climate change from greenhouse gas emissions and the adverse consequences that flow from these emissions is the principal justification for government support of the nuclear energy option."[13] But environmental groups generally contend that the nuclear accident, waste, and weapons proliferation risks posed by nuclear power outweigh any GHG benefits. The large construction expenditures required by commercial reactors, they contend, would yield greater GHG reductions if used for energy efficiency and renewable generation. Finally, they note that nuclear power, while not directly emitting greenhouse gases, produces indirect emissions through the nuclear fuel cycle and during plant construction.

Another key factor behind the renewed interest in nuclear power is the improved performance of existing reactors. U.S. commercial reactors generated electricity at an average of 90% of their total capacity in 2008,[14] after averaging around 75% in the mid-1990s and around 65% in the mid-1980s. Worldwide performance has seen similar improvement.[15] The improved operation of nuclear power plants has helped drive down the cost of nuclear-generated electricity. Average U.S. reactor operations and maintenance costs (including fuel but excluding capital costs) dropped steadily from a high of about 3.5 cents/kilowatt-hour (kwh) in 1987 to below 2 cents/kwh in 2001 (in 2001 dollars).[16] U.S. nuclear plant operating costs during 2006-2008 averaged about 1.9 cents/kwh.[17]

Nuclear interest has been further increased in the United States by incentives in the Energy Policy Act of 2005 (P.L. 109-58). The law provides a nuclear energy production tax credit for up to 6,000 megawatts of new nuclear capacity, compensation for regulatory delays for the first six new reactors, and federal loan guarantees for nuclear power and other advanced energy technologies. Under certain baseline assumptions, the tax credits and loan guarantees could determine whether new U.S. nuclear plants would be economically viable.[18]

U.S. electric utilities and other companies during the past two years have announced plans to submit license applications to the Nuclear Regulatory Commission (NRC) for about 30 new commercial reactors (see Table 1.) NRC has issued "early site permits"—which resolve site-related issues for possible future reactor construction—at locations in Illinois, Mississippi, Virginia, and Georgia. The Tennessee Valley Authority in 2007 restarted construction of its long-delayed Watts Bar 2 reactor, which had been ordered in 1970. But despite that flurry of activity, no new reactor orders have been formally placed. No reactors have been ordered in the United States since 1978, and all orders after 1973 were subsequently cancelled.

The recent nuclear plant license applicants have not committed to building new reactors if the licenses are approved, but the sponsors of four of the proposed nuclear projects have signed preliminary engineering, procurement, and construction (EPC) contracts. President Obama announced a conditional loan guarantee on February 16, 2010, for two proposed reactors at Georgia's Vogtle nuclear power plant, which had previously received an NRC early site permit. On the other hand, Entergy suspended further license review of its planned GE ESBWR reactors at River Bend, LA, and Grand Gulf, MS, and Dominion is seeking other potential vendors for its planned ESBWR at North Anna, VA, although it is continuing with the licensing process. AmerenUE has suspended construction plans for a new reactor at its Callaway plant, while also continuing with NRC licensing.

Table 1. Announced Nuclear Plant License Applications

Announced Applicant	Site	Planned Application	Reactor Type	Units	Status
Alternate Energy	Hammett (ID)	2009	Areva EPR	1	
AmerenUE	Callaway (MO)	Submitted 7/24/08	Areva EPR	1	Construction plans suspended 4/23/09; NRC license review suspended 6/23/09
Amarillo Power	Near Amarillo (TX)	2009	Areva EPR	2	
Dominion	North Anna (VA)	Submitted 11/27/07	GE ESBWR	1	Other reactor vendors being considered 1/9/09
DTE Energy	Fermi (MI)	Submitted 9/18/08	GE ESBWR	1	
Duke Energy	Cherokee (SC)	Submitted 12/13/07	Westing.house AP1000	2	
Entergy	River Bend (LA)	Submitted 9/25/08	Not specified	1	Licensing suspended 1/9/09
Luminant Power (formerly TXU)	Comanche Peak (TX)	Submitted 9/19/08	Mitsubishi US-APWR	2	
FPL	Turkey Point (FL)	Submitted 6/30/09	Westinghouse AP1000	2	
NRG Energy	South Texas Project	Submitted 9/20/07	GE ABWR	2	EPC contract signed with Toshiba 2/12/09
NuStart	Grand Gulf (MS), Entergy	Submitted 2/27/08	Not specified	1	Licensing suspended Jan. 9, 2009
	Bellefonte (AL), TVA	Submitted 10/30/07	Westinghouse AP1000	2	NuStart announced shift of lead unit to Vogtle 4/30/09
PPL	Bell Bend (PA)	Submitted 10/10/08	Areva EPR	1	
Progress Energy	Harris (NC)	Submitted 2/19/08	Westinghouse AP1000	2	EPC contract signed 1/5/09
	Levy County (FL)	Submitted 7/30/08	Westinghouse AP1000	2	

Table 1. (Continued)

Announced Applicant	Site	Planned Application	Reactor Type	Units	Status
SCE&G	Summer (SC)	Submitted 3/31/08	Westinghouse AP1000	2	EPC contract signed 5/27/08
Southern	Vogtle (GA)	Submitted 3/31/08	Westinghouse AP1000	2	EPC contract signed 4/8/08; Vogtle to be NuStart lead unit; NRC granted early site permit 8/26/09; conditional DOE loan guarantee announced 2/16/10
UniStar (Constellation Energy and EDF)	Calvert Cliffs (MD)	Submitted 7/13/07 (Part 1), 3/13/08 (Part 2)	Areva EPR	1	
	Nine Mile Point (NY)	Submitted 9/30/08	Areva EPR	1	Licensing suspended 12/1/09
Total Units				**29**	

Sources: NRC, *Nucleonics Week*, *Nuclear News*, Nuclear Energy Institute, company news releases.

New reactors are on order elsewhere in the world, and several non-nuclear countries have announced that they are considering the nuclear option. As Figure 2 shows, the vast majority of reactors currently under construction are in Asia, with only a handful in the rest of the world.

Despite the recent positive developments for nuclear power, much uncertainty still remains about its prospects. Construction costs for new nuclear power plants—which were probably the dominant factor in halting the first round of nuclear expansion—continue to loom as a potential insurmountable obstacle to renewed nuclear power growth. Average U.S. nuclear plant construction costs more than doubled from 1971 to 1978, according to the Office of Technology Assessment, and nearly doubled again by the mid-1980s, not including interest accrued during construction.[19] Including interest, many U.S. nuclear plants proved to be grossly uneconomic, often with capital costs totaling more than $3,000 per kilowatt of capacity in 2000 dollars,[20] and relying on the utility regulatory system to recover their costs.

Major reactor vendors, such as General Electric and Westinghouse, have contended that new designs and construction methods will cut costs of future reactors considerably. However, recent projected costs for new reactors average $3,900 per kilowatt, excluding interest, making them potentially more expensive than the previous generation of reactors.[21] Capital costs of competing power generation technologies, particularly coal, have also risen in recent years.

New U.S. nuclear plants may be helped by a new NRC licensing process that is intended to avoid some of the regulatory problems that delayed completion of some reactors in the past. Current reactor applications listed in Table 1 are being evaluated by NRC under the new system, but no licenses under the new system have yet been issued.

Many other important factors in the future of nuclear power are similarly uncertain. Prices of competing fuels, particularly natural gas, have risen recently but have been volatile in the recent past. If fossil fuel prices become depressed for a sustained period, as in the late 1980s through the 1990s, support for nuclear power as an alternative energy source could again be undermined. Major accidents, such as Three Mile Island and Chernobyl, would almost certainly diminish public support for nuclear power. Disposal of high-level nuclear waste, which reprocessing or recycling is intended to address, will continue to generate controversy as governments attempt to develop permanent underground repositories—none of which are yet operating.

Worldwide Nuclear Power Status

Operating commercial nuclear reactors around the world currently total 436, with total installed electric generating capacity of 372 gigawatts.[22] More than 80% of world nuclear capacity is in member nations of the Organization for Economic Cooperation and Development (OECD), while slightly more than 10% is in Russia and other former nations of the Soviet bloc. The remainder, about 5%, is in developing countries such as China and India. Nuclear power supplied 27.3% of electricity generated in OECD countries and 5.2% in non-OECD countries in 2006.[23]

Unlike the United States, where only one long-deferred reactor is currently under construction, much of the rest of the world has continued building nuclear plants, although at a modest pace. Since 1996, about 40 commercial reactors have started up, an average of about four per year. About 30 reactors were permanently closed during that period, although many of them were smaller than the newly started reactors.[24]

Current reactor construction is dominated by Asia, as shown by Figure 2. Of the 53 reactors now under construction around the world, 35 are in Asia, while 13 are in Europe, four in the Americas, and one in the Middle East (Iran). Planned or proposed nuclear power plants show a similar trend. Of the 469 planned or proposed reactors in Figure 2, well more than half (254) are in Asia, while 121 are in Europe, 47 in the Americas, and 20 in the Middle East. South Africa has proposed up to 27 new reactors.

The renewed worldwide interest in nuclear power has led to a possible expansion of the technology to currently non-nuclear nations. Nine of the countries that are currently building or formally planning reactor projects—Belarus, Egypt, Indonesia, Iran, Kazakhstan, Thailand, Turkey, the United Arab Emirates, and Vietnam—have never operated nuclear power plants. Several other non-nuclear power countries have proposed building power reactors, including Bangladesh, Israel, and Poland.[25]

Nuclear Fuel Services Market

The possible upsurge in worldwide nuclear power plant construction has focused new attention on nuclear fuel production. Chronic worldwide overcapacity in all phases of the nuclear fuel cycle appears to be ending, evidenced by higher prices for uranium and enrichment services in recent years. The tightening supplies have sparked plans for new fuel cycle facilities around the world and also renewed concerns about controls over the spread of nuclear fuel technology.

The nuclear fuel cycle begins with mining uranium ore, and upgrading it to yellowcake. Because naturally occurring uranium lacks sufficient fissile ^{235}U to make fuel for commercial light-water reactors, the concentration of ^{235}U must be increased several times above its natural level of 0.7% in a uranium enrichment plant. A nuclear power plant operator or utility typically purchases yellowcake and contracts for its conversion to uranium hexafluoride, then enrichment, and finally fabrication into fuel elements (Figure 1). Commercial enrichment services are available in the United States, Europe, Russia, and Japan. Fuel fabrication services are even more widely available. While waiting for conversion, the yellowcake remains a fungible commodity that can be consigned by the reactor operator to any conversion plant and the product sent to any enrichment plant (within trade restrictions between countries).[26] The sale of yellowcake had been informal, until recently when it moved to a more formal commodity transaction basis. The various stages of the nuclear fuel cycle are described below.

Yellowcake

Conventionally mined uranium ore (open-pit and underground) is milled, then acid-leached to extract uranium oxide. The extract is then filtered, dried, and packaged as uranium yellowcake for shipment to a conversion plant. In situ leaching avoids the mechanical mining steps by directly injecting solvents into the ore body through wells drilled from the surface. The dissolved uranium is pumped to the surface, where the uranium oxide is similarly processed into yellowcake for shipment.

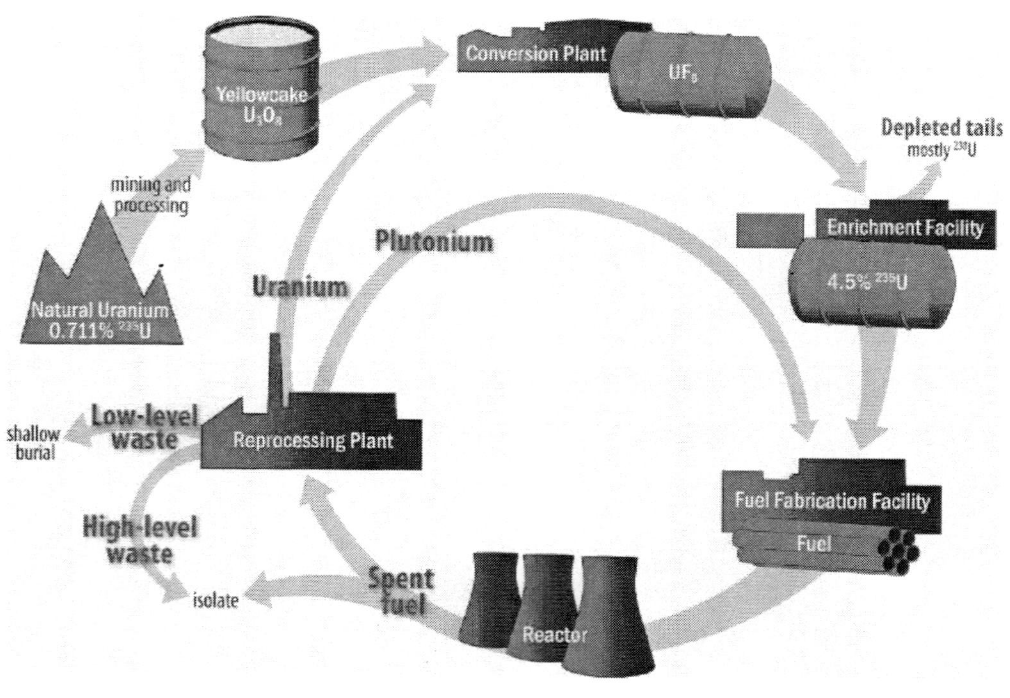

Figure 1. The Conceptual Nuclear Fuel Cycle

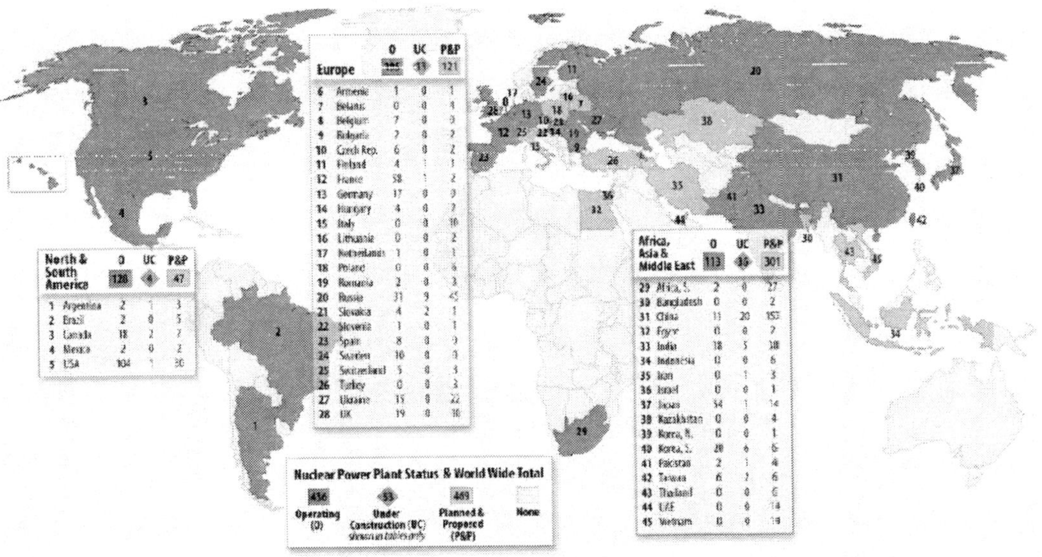

Source: World Nuclear Association, http://www.world-nuclear.org/info/reactors.html

Figure 2. World Wide Nuclear Power Plants Operating, under Construction, and Planned

U.S. uranium reserves are located in Arizona, Colorado, Nebraska, New Mexico, Texas, Utah, Washington, and Wyoming. According to the Energy Information Administration (EIA), 10 underground mines and six in-situ mines were operating in the United States in 2008, four more than the previous year. EIA reports 53 million pounds of U3O8 were purchased for U.S. nuclear power reactors in 2008, of which 14% was U.S. origin.[27] The balance was made up in part by imports and downblended highly enriched uranium (HEU), as discussed further below.

A typical 1,000 MW light water reactor fuel load may require converting and enriching nearly 800,000 lbs of uranium “yellowcake” (U3O8). Approximately 51,800 metric tons (114 million lbs) of yellowcake was produced worldwide in 2008. Annual worldwide uranium demand is estimated to be about 78,500 metric tons of U3O8. Most of the difference between annual production and demand is covered by sales from former military uranium stockpiles.[28] The International Atomic Energy Agency (IAEA) projects that the demand for uranium will begin to exceed supply after 2010, and by as much as 10,000 metric tons by 2020.[29] IAEA believes that the shortfall could be made up by downblending more HEU released from weapons stockpiles.

Unlike gold or oil commodities, uranium yellowcake had not been offered through a formal market exchange until quite recently. Uranium price indicators had been developed by a small number of private business organizations, such as the World Nuclear Fuel Market (WNFM) and the Ux Consulting Company (UxC), that independently monitor uranium market activities, including offers, bids, and transactions. The price indicators are owned by and proprietary to the businesses that has developed them.

NAC International (now a USEC Inc. subsidiary) established the World Nuclear Fuel Market (WNFM) to provide uranium price information in 1974. The WNFM membership comprises 79 companies representing 18 countries.[30] The WNFM provides the uranium price information system (UPIS) for both Western and Russian yellowcake contract prices.[31] A quarterly UPIS report presents aggregated information based on actual uranium contract price data provided by the 19 UPIS subscribers.[32]

The UxC pricing index has been utilized by major nuclear fuel market participants, the federal government, and private business. The UxC yellowcake price was one of only two weekly uranium price indicators that were accepted by the uranium industry, as witnessed by their inclusion in most "market price" sales contracts; that is, sales contracts with pricing provisions that call for the future uranium delivery price to be equal to the market price at or around the time of delivery.

In April 2007, the New York Mercantile Exchange (NYMEX) announced that it had partnered with the UxC to provide financially settled on- and off-exchange traded uranium futures contracts.[33] A NYMEX uranium futures contract's final settlement price is based on the UxC pricing index for yellowcake. Uranium futures contracts are available for trading on Chicago Mercantile Exchange Globex, and for clearing on NYMEX ClearPort.[34] The size of each contract is 250 lbs, and prices are quoted in U.S. currency. The final settlement price is the spot month-end price published by UxC. A standard contract for trading physical uranium is currently under development.[35]

Uranium is typically mined outside the countries that use it. Nearly 60% of the world's production in 2008 came from Canada, Kazakhstan, and Australia, while more than half of the world's commercial reactors are in the United States, France, and Japan.[36] But security of uranium supply, while always an underlying policy concern, has rarely been a real problem, because production vastly outstripped demand during the first three decades of the commercial nuclear power era—until about the mid-1980s.[37] As a result, a huge overhang of military and civilian stockpiles of uranium helped maintain a worldwide buyers' market.

Since the mid-1980s, however, world nuclear fuel requirements continued to rise while uranium exploration and production fell. By 2000, as U.S. spot-market prices hit bottom (at about $7 per pound), the western world's nuclear fuel requirements were twice the level of production. At that point, commercial stockpiles had been drawn down enough to begin putting pressure on U.S. spot prices, which rose slightly through 2003 and then dramatically (above $75 per pound) in 2007, before falling back to the $50 range in 2009 and early 2010. The spot price represents about 20% of the market but provides an indicator of future contracts, which usually run three to seven years.[38]

Despite low worldwide exploration expenditures since the mid-1980s caused by oversupply and low prices, estimated uranium resources have trended upward over the long term. As a result, according to the OECD Nuclear Energy Agency (NEA), known conventional resources have averaged 45 years of supply during the past 20 years, despite steadily increasing annual world uranium requirements, currently about 70,000 metric tons. "Taken together the lessons of the past provide confidence that uranium resources will remain adequate to meet projected demands even were requirements to significantly increase," according to NEA.[39]

Conversion

In the conversion process, the yellowcake is purified, chemically reacted with hydrofluoric acid to form uranium hexafluoride (UF_6) gas, and then transferred into cylinders, where it cools and condenses to a solid. Uranium hexafluoride contains two isotopes of uranium—heavier ^{238}U and lighter fissionable ^{235}U, which makes up ~0.7% of uranium by weight. The annual U.S. demand for yellowcake conversion is approximately 22,000 metric tons uranium (MTU). After conversion, the uranium hexaflouride is ready for enrichment.

Five commercial conversion companies operate worldwide—in the United States, Canada, France, the United Kingdom, and Russia (Table 2). ConverDyn in Metropolis, IL, the only conversion plant operating in the United States, produces 14,000 MTU annually.

Enrichment

For use as fuel in light water reactors, uranium must be enriched in the isotope ^{235}U above its natural concentration of 0.7%. By heating UF6 to turn it into a gas, the enrichment process can take advantage of the slight difference in atomic mass between 235U and 238U. The typical enrichment process requires about 10 lbs of uranium U3O8 to produce 1 lb of low-enriched uranium hexafluoride (UF6) product, with a 235U concentration of 3-5%.

About 90% of the world's reactors (all except heavy water reactors) require enriched uranium fuel. More than 90% of those uranium enrichment requirements are supplied by facilities in the United States (including diluted weapons material), Russia, France, Great Britain, Germany, and the Netherlands. The remainder comes from Japan, China, and Brazil. Thirty-one countries currently operate commercial nuclear power plants. Most countries, therefore, rely on enrichment services outside their borders. An enrichment plant to serve a country with only a few reactors would appear economically nonviable, given that a single large enrichment plant can supply up to 25% of the world market (currently estimated at 45 million separative work units, or SWUs).[40]

Commercial uranium enrichment employs either gaseous diffusion or high speed centrifuges. In gaseous diffusion, a thin semiporous barrier holds back more of the heavier ^{238}U than the lighter ^{235}U. A series of cascading diffusers successively increases the ^{235}U concentration. Centrifuge enrichment spins the uranium hexafluoride gas at ultra-high speeds to separate the lighter ^{235}U. A series of cascading centrifuges successively enriches the gas in ^{235}U. Final enrichment will vary depending on the requirements of a specific reactor, normally up to about 5%.

Gaseous diffusion technology was first developed in the United States and later adopted by France and Britain. It is more energy-intensive than the newer centrifuge enrichment process. However, the legacy gaseous diffusion plants currently operating in the United States and France have higher capacities than the newer centrifuge enrichment plants.

Table 2. Commercial UF6 Conversion Facilities (metric tons uranium/year)

Country	Company	Facility	Capacity
Canada	Cameco	Port Hope	12,500
China	CNCC	Lanzhou	1,500
France	Comurhex	Peirrelatte 1	14,000
		Peirrelatte 2	350
Russian	Minatom	Angarsk	20,000
Federation		Tomsk	10,000
U.K.	BNFL	Springfields Line 4	6,000
U.S.	Converdyn	Metropolis	14,000

Source: IAEA Country Nuclear Fuel Cycle Profiles, 2nd ed.

Uranium enrichment services are sold in kilograms (kg) or metric tons (1,000 kg) separative work units (SWU), which is a measure of the amount of work needed (in the thermodynamic sense) to enhance the ^{235}U concentration. The number of SWUs required to produce fuel depends on several factors: the quantity of fuel required, level of enrichment required, the initial enrichment of the feed (0.711% in the case of natural uranium), and the "tails assay," which is the ^{235}U concentration remaining in the depleted processing stream. For example, to produce 1 kg of uranium enriched to 3% ^{235}U, at a tails assay of 0.2 ^{235}U, 4.3 kg-SWU are used to process 5.5 kg of natural uranium.[41] The price of yellowcake is an important factor in enrichment demand. Under high price conditions, it may be economically preferable to expend more SWUs enriching a lesser quantity of yellowcake, thus leaving a lower tails assay.

Nuclear plant operators can buy uranium yellowcake and have it converted and enriched, or buy low-enriched uranium (LEU). Commercial enrichment services are offered by a number of international sources (Table 3), with worldwide annual capacity of 47,855 metric tons SWU. In 2008, U.S. nuclear plant operators contracted five companies worldwide to provide 12.6 million kilogram SWU (12,600 metric tons SWU, of which 15% was provided in the United States.[42]

The U.S. DOE had operated gaseous diffusion enrichment plants in Oak Ridge, TN, Paducah, KY, and Portsmouth, OH, to produce high-enriched uranium used in the nuclear weapons program. The plants later produced low-enriched uranium for commercial nuclear power around the world, although production at the Oak Ridge K-25 enrichment site ceased in 1985. The Energy Policy Act of 1992 established the United States Enrichment Corporation (USEC) as a government-owned corporation to take over DOE's uranium enrichment services business. The corporation was privatized as USEC Inc. in 1998. In 2001, USEC ceased uranium enrichment operations in Portsmouth and consolidated operations in Paducah. The Paducah gaseous diffusion plant is the only operating enrichment facility in the United States. In 2004, USEC announced plans to build the American Centrifuge Plant on the site of the Portsmouth, OH, gaseous diffusion plant. The new gas centrifuge enrichment plant is to have 11,500 centrifuges with an annual capacity of 3.8 million SWU.[43] Construction of the plant was suspended in August 2009, with restart dependent on DOE approval of a loan guarantee.[44] USEC currently supplies approximately 51% of the U.S. demand for enrichment services, mostly with blended-down Russian HEU, as discussed below.

Urenco, a joint Dutch, German, and British enrichment consortium, was set up in the 1970s following the signing of the Treaty of Almelo. Urenco operates enrichment plants in Germany, the Netherlands, and the United Kingdom to supply customers in Europe, North America, and East Asia. Its U.S. affiliate, Louisiana Energy Services, has begun constructing the gas centrifuge National Enrichment Facility (NEF) in New Mexico.[45] NEF is expected to produce 3 million SWUs annually when it reaches full operational capacity in 2013—meeting approximately 25% of the current U.S. demand.[46] In 2006, Urenco estimated that it provided around 23% of the world market share in enrichment services.

Areva operates the Eurodif gaseous diffusion production plant (located on the Tricastin nuclear site in France) to enrich uranium for some 100 nuclear reactors in France and throughout the world.[47] Areva provides toll conversion services and uranium yellowcake through its subsidiary Comurhex. Areva applied to NRC at the end of 2008 for a license to build and operate a gas centrifuge enrichment plant in Idaho.[48]

Table 3. Operating Commercial Uranium Enrichment Facilities (metric tons SWU/year)

Facility Name	Country	Process	Capacity
Paducah Gaseous Diffusion	United States	Gaseous Diffusion	11,300
Eurodif (Georges Besse)	France	Gaseous Diffusion	10,800
Ekaterinburg (Sverdlovsk-44)	Russian Federation	Centrifuge	7,000
Siberian Chemical Combine (Seversk)	Russian Federation	Centrifuge (downblended)	4,000
Urenco Capenhurst	United Kingdom	Centrifuge	4,000
Krasnoyarsk	Russian Federation	Centrifuge	3,000
Urenco Nederland	Netherlands	Centrifuge	2,900
Urenco Deutschland	Germany	Centrifuge	1,800
Rokkasho Uranium Enrichment Plant	Japan	Centrifuge	1,050
Angarsk	Russian Federation	Centrifuge	1,000
Lanzhou 2	China	Centrifuge	500
Shaanxi Uranium Enrichment Plant	China	Centrifuge	500
Kahuta	Pakistan	Centrifuge	5
Total			**47,855**

Source: International Atomic Energy Agency, Nuclear Fuel Cycle Information System.

Under the 1993 U.S.-Russian Federation Megatons to Megawatts program, highly enriched uranium from dismantled Russian nuclear warheads is converted into low-enriched uranium fuel for use in commercial U.S. nuclear power plants.[49] The HEU Agreement, as it is known, provides for the purchase over 20 years of 500 metric tons highly enriched uranium downblended to commercial grade low-enriched uranium (delivered as UF_6). The agreement provides about 46% of the current U.S. demand for enrichment.

The world uranium enrichment industry is currently undergoing a technological transformation from gaseous diffusion to centrifuges, primarily because centrifuges need only a fraction of the energy required by gaseous diffusion. In 1996, 57% of the world's commercial enrichment came from gaseous diffusion plants, a level that dropped to 35% in 2006. As noted above, the United States' only currently operating enrichment facility, in Paducah, KY, is planned to be replaced with a centrifuge plant in Portsmouth, OH. The world's only other operating gaseous diffusion plant, at Areva's Tricastin site in France, is to be replaced by a centrifuge plant by around 2012.[50]

Fuel Fabrication

Like enrichment, fuel fabrication is a specialized service rather than a commodity transaction. The now low-enriched uranium (UF_6) undergoes one final process, converting to uranium dioxide (UO_2), before the final stage of fuel fabrication. It is then sintered into pellets and loaded into zirconium alloy tubes (fuel rods) about 12-15 feet long and half an inch in diameter. The fuel rods are bundled into fuel assemblies, which vary from fewer than 100 to as many as 300 rods apiece.

Fuel fabrication services are offered by 16 suppliers operating in 18 countries at around 34 facilities. In 2002, IAEA estimated that worldwide fabrication capacity of 19,000 tons (fuel assemblies and elements) exceeded the demand by 53%.[51] The oversupply had existed for many years, and, as a consequence, facilities were shut down and ownership was consolidated.

Essentially all U.S. fabrication demand is met by three companies providing fabrication service at four facilities: Framatome ANP Inc. in Lynchburg, VA, and Richland, WA; Global Nuclear Fuel in Wilmington, NC; and Westinghouse Electric in Columbia, SC. About 30 other nuclear fuel fabrication facilities are in operation elsewhere in the world.[52]

Final Stages of the Fuel Cycle

The final stages of the nuclear fuel cycle take place after nuclear fuel assemblies have been loaded into a reactor. In the reactor, the uranium 235 (^{235}U) splits, or fissions, releasing energy, neutrons, and fission products (highly radioactive fragments of ^{235}U nuclei). The neutrons may cause other ^{235}U nuclei to fission, creating a nuclear chain reaction. Some neutrons are also absorbed by ^{238}U nuclei to create plutonium 239 (^{239}Pu), which itself may then fission.

After several years in the reactor, fuel assemblies will build up too many neutron-absorbing fission products and become too depleted in fissile ^{235}U to efficiently sustain a nuclear chain reaction. At that point, the assemblies are considered spent nuclear fuel and removed from the reactor. Spent fuel typically contains about 1% ^{235}U, 1% plutonium, 4% fission products and other radioactive waste, and the remainder ^{238}U.

The last stage of the fuel cycle, after spent fuel is removed from a reactor, has proved highly contentious. One option is to directly dispose of spent fuel in a deep geologic repository to isolate it for the hundreds of thousands of years that it may remain hazardous. The other option is to reprocess the spent fuel to separate the uranium and plutonium for use in new fuel. Supporters of reprocessing, or recycling, contend that it could greatly reduce the volume and longevity of nuclear waste while vastly expanding the amount of energy extracted from the world's uranium resources. Opponents contend that commercial use of separated plutonium—a key material in nuclear weapons as well as reactor fuel—poses a nuclear weapons proliferation threat.

Commercial-scale spent fuel reprocessing is currently conducted in France, Britain, and Russia. The ^{239}Pu they produce is blended with uranium to make mixed-oxide (MOX) fuel, in which the ^{239}Pu largely substitutes for ^{235}U. Two French reprocessing plants at La Hague can each reprocess up to 800 metric tons of spent fuel per year, while Britain's THORP facility at Sellafield has a capacity of 900 metric tons per year. Russia has a 400-ton plant at Ozersk, and Japan is building an 800-ton plant at Rokkasho to succeed a 90-ton demonstration facility at Tokai Mura. Britain and France also have older plants to reprocess gas-cooled reactor fuel, and India has a 275-ton plant.[53] About 200 metric tons of MOX fuel is used annually, about 2% of new nuclear fuel,[54] equivalent to about 2,000 metric tons of mined uranium.[55]

However, the benefits of reprocessing spent fuel to make MOX fuel for today's nuclear power plants are modest. Existing commercial light water reactors use ordinary water to slow down, or "moderate," the neutrons released by the fission process. The relatively slow (thermal) neutrons are highly efficient in causing fission in certain isotopes of heavy elements, such as ^{235}U and ^{239}Pu.[56] Therefore, lower quantities of those isotopes are needed in nuclear fuel to sustain a nuclear chain reaction. The downside is that thermal neutrons cannot efficiently induce fission in more than a few specific isotopes. In today's commercial reactors, therefore, the buildup of non-fissile plutonium and other isotopes sharply limits the number of reprocessing cycles before the recycled fuel can no longer sustain a nuclear chain reaction and must be stored or disposed of.

In contrast, "fast" neutrons, which have not been moderated, are less effective in inducing fission than thermal neutrons but can induce fission in all actinides, including all plutonium isotopes. Therefore, nuclear fuel for a fast reactor must have a higher proportion of fissionable isotopes than a thermal reactor to sustain a chain reaction, but a larger number of different isotopes can constitute that fissionable proportion.

A fast reactor's ability to fission all actinides (actinium and heavier elements), makes it theoretically possible to repeatedly separate those materials from spent fuel and feed them back into the reactor until they are entirely fissioned. Fast reactors are also ideal for "breeding" the maximum amount of ^{239}Pu from ^{238}U, eventually converting virtually all of natural uranium to useable nuclear fuel.[57] Current reprocessing programs are generally viewed by their proponents as interim steps toward a commercial nuclear fuel cycle based on fast reactors.

Waste Disposal and Energy Security

Reprocessing of spent fuel from fast breeder reactors has long been the ultimate goal of nuclear power supporters. As noted above, fast reactors (operated either as breeders or non-breeders) can largely eliminate plutonium from nuclear waste and greatly extend uranium supplies. But opponents contend that such potential benefits are not worth the costs and nonproliferation risks.

Removing uranium from spent nuclear fuel through reprocessing would eliminate most of the volume of radioactive material requiring disposal in a deep geologic repository. In addition, the removal of plutonium and conversion to shorter-lived fission products would eliminate most of the long-term (post-1,000 years) radioactivity in nuclear waste. But the waste resulting from reprocessing would have nearly the same short-term radioactivity and heat as the original spent fuel, because the reprocessing waste consists primarily of fission products, which generate most of the radioactivity and heat in spent fuel. Because heat is the main limiting factor on repository capacity, conventional reprocessing would not provide major disposal benefits in the near term.

To address that problem, various proposals have been made to further separate the primary heat-generating fission products—cesium 137 and strontium 90—from high-level waste for separate storage and decay over several hundred years. Such a process could greatly increase repository capacity, although it would require an alternative secure storage system for the cesium and strontium for hundreds of years.

Energy security has been a primary driving force behind the development of nuclear energy, particularly in countries such as France and Japan that have few natural energy resources. Recent cutoffs in oil and gas around the world have underscored the instability of oil and gas supply, which could be mitigated by nuclear energy. For example, in 2006, a natural gas price dispute between Russia and Ukraine resulted in a temporary cutoff of natural gas to Western and Central Europe; in 2007, price disputes between Russia and Azerbaijan and Belarus caused a temporary cutoff in oil to Russia from Azerbaijan and in oil from Russia to Germany, Poland, and Slovakia.

Moreover, temporary production shutdowns in the Gulf of Mexico and the Trans-Alaskan pipeline, instability in Nigeria, and nationalization of oil and gas fields in Bolivia in 2006 have all raised concerns about oil and gas supplies and worldwide price volatility. Relative to gas and oil, the ability to stockpile uranium is widely seen as offering greater assurances of weathering potential cutoffs.

Worldwide uranium resources are generally considered to be sufficient for at least several decades. Uranium supply is highly diversified, with uranium mining spread across the globe, while uranium conversion, enrichment, and fuel fabrication are more concentrated in a handful of countries. But because most reactors around the world rely at least partly on foreign sources of uranium and nuclear fuel services, nuclear reactors nearly everywhere face some level of supply vulnerability. To mitigate such concern, countries such as China, India, and Japan are seeking to secure long-term uranium contracts to support nuclear expansion goals. Efforts are underway to establish an international nuclear fuel bank to provide greater certainty in fuel supplies, as discussed in the next section.

Ultimately, only the development of breeder reactors and reprocessing could provide complete nuclear energy independence. This remains the long-term goal of resource-poor France and Japan, and Russia as well, although their research and development programs have faced numerous obstacles and schedule slowdowns.

PROPOSALS ON THE FUEL CYCLE[58]

Proposals addressing access to the full nuclear fuel cycle have ranged from seeking a formal commitment to forswear enrichment and reprocessing technology, to a de facto approach in which a state does not operate fuel cycle facilities but makes no explicit commitment to give them up, to no restrictions at all. Current proposals generally aim to persuade countries not to develop their own fuel production capabilities by providing economically attractive alternatives that allay concerns about politically motivated interruption to fuel supply. Most proposals focus on this front-end problem, dealing with fuel supply and production issues. Ultimately, these proposals are aimed at preventing an increase in the number of states that would be capable of producing weapons-usable nuclear material.

The main proposals under discussion are outlined here in categories—those addressing the full fuel cycle, those addressing the "front-end" or assurance of fuel supply issues (including fuel banks and multilateral enrichment services), and those focusing on the "back-end" or waste disposal solutions.

Comprehensive Proposals

El Baradei Proposal

IAEA Director General Mohamed El Baradei in 2003 proposed a three-pronged approach to limiting the processing of weapon-usable material (separated plutonium and high-enriched uranium) in civilian nuclear fuel cycles.[59] First, he would place all enrichment and reprocessing facilities under multinational control. Second, he would develop new nuclear technologies that would not produce weapons-usable fissile material—in other words, "the holy grail" of a proliferation-resistant fuel cycle. In his October 2003 article in the *Economist* where he laid out these ideas, El Baradei maintained, "This is not a futuristic dream; much of the technology for proliferation-resistant nuclear-energy systems has already been developed or is actively being researched." Third, El Baradei proposed considering "multinational approaches to the management and disposal of spent fuel and radioactive waste." El Baradei did not place any nonproliferation requirements on participation, but instead suggested that the system "should be inclusive; nuclear-weapon states, non-nuclear-weapon states, and those outside the current nonproliferation regime should all have a seat at the table." Further, he noted that a future system should achieve full parity among all states under a new security structure that does not depend on nuclear weapons or nuclear deterrence.

IAEA Experts Group/INFCIRC/640

In February 2005, an Expert Group commissioned by IAEA Director General El Baradei presented a report, "Multilateral Approaches to the Nuclear Fuel Cycle."[60] The Expert Group studied several possible approaches to securing the operation of proliferation-sensitive nuclear fuel cycle activities (uranium enrichment, reprocessing and spent fuel disposal, and storage of spent fuel) and analyzed the incentives and disincentives for states to participate. The report reviewed relevant past and present experience. The Group's suggested approaches included the following:

- Reinforce existing market mechanisms by providing additional supply guarantees by suppliers and/or the IAEA (fuel bank).
- Convert existing facilities to multinational facilities.
- Create co-managed, jointly owned facilities.

The Group concluded that "in reality, countries will enter into multilateral arrangements according to the economic and political incentives and disincentives offered by these arrangements."[61] The report noted that no legal framework existed for requiring states to join supply assurance arrangements.

In September 2006, the IAEA sponsored a conference entitled "New Framework for the Utilization of Nuclear Energy in the 21st Century: Assurances of Supply and Non-Proliferation," which addressed proposals to provide fuel assurances. The IAEA presented a report on fuel assurance options at the June 2007 Board of Governors meeting analyzing the various proposals put forth to date.[62] A potential framework for nuclear supply assurances could have three stages: (1) existing market arrangements; (2) back-up commitments by suppliers in case of a politically motivated interruption of supply if nonproliferation criteria are met; (3) a physical LEU material reserve.[63] The report emphasizes that participation in these arrangements should be voluntary, that progress on this question will be incremental and that many options should be explored to give consumer states sufficient choices to meet their needs. The report is still under discussion by IAEA Board members.

President Bush's 2004 Proposal

In a speech at the National Defense University on February 11, 2004, President Bush said the world needed to "close a loophole" in the NPT that allows states to legally acquire the technology to produce nuclear material which could be used for a clandestine weapons program. To remedy this, he proposed that the 40 members of the Nuclear Suppliers Group (NSG) should "refuse to sell enrichment and reprocessing equipment and technologies to any state that does not already possess full-scale, functioning enrichment and reprocessing plants."[64] President Bush also called on the world's leading nuclear fuel services exporters to "ensure that states have reliable access at reasonable cost to fuel for civilian reactors, so long as those states renounce enrichment and reprocessing."

President Bush's 2004 proposal was the only one that calls for countries to explicitly "renounce" pursuit of enrichment or reprocessing technologies in exchange for reliable access to nuclear fuel, and proved controversial. Many non-nuclear-weapon states saw this as an attempt to limit their inalienable right to the use of peaceful nuclear energy under Article IV of the NPT and are not willing to consider limits on peaceful nuclear technologies until more progress on nuclear disarmament has been made.

Fundamental questions about this original formulation were raised. For example, who is included in the group of supplier states and how is "full-scale, functioning plants" defined? Would Iran's enrichment program qualify today, even if it did not back in 2004? And what about non-NPT states with the full fuel cycle as part of their weapons programs? Also, how would related technologies be treated? For example, would restrictions also apply to post-irradiation experiments on spent nuclear fuel, which yield significant data about reactor operations, but can also contribute to knowledge about reprocessing for weapons purposes? Since 2004, delay in defining these terms appears to have provided an incentive for some states, such as Canada, South Africa, Argentina, and Australia, to expedite their pursuit of a full operational enrichment capability so as not to be excluded when and if such a division between fuel cycle haves and have-nots is made.

Russia's "Global Nuclear Power Infrastructure"

In January 2006, Russian President Vladimir Putin proposed the Global Nuclear Power Infrastructure initiative that would include four kinds of cooperation: creation of international uranium-enrichment centers (IUECs), international centers for reprocessing and storing spent nuclear fuel, international centers for training and certifying nuclear power plant staff, and an international research effort on proliferation-resistant nuclear energy technology. The international fuel cycle centers would be under joint ownership and co-management. They would be commercial joint ventures (that is, no state financing), with advisory boards consisting of government, industry, and IAEA professionals. The IAEA would not have a vote on these boards, but would play an advisory role, while also certifying the fuel provision commitments.

Recipient countries under Putin's proposal would receive fuel cycle services, but access to sensitive technology would stay in the hands of the supplier state. Russia has offered a similar arrangement to Iran—to jointly enrich uranium on Russian territory. Iran has not yet accepted this offer, but it is still part of ongoing negotiations with Iran over its nuclear program. Russia has also made the return of spent fuel from the Bushehr nuclear plant it is constructing in Iran a condition of supply, so that no plutonium can be extracted from the spent fuel.[65] To date, progress has been made is establishing the IUEC and in establishing an LEU reserve at Angarsk (see below). Russia has also announced creation of a center of excellence for nuclear personnel at Obninsk, and actively participates in joint research efforts on fast reactors such as the International Project on Innovative Nuclear Reactors and Fuel Cycles (INPRO).

Assurance of Fuel Supply: Supplier Guarantees

The following proposals focus on back-up fuel supply assurances designed to complement, but not impact or supplant, the commercial uranium market.

Six Country Concept

In May 2006, six governments—France, Germany, the Netherlands, Russia, the United Kingdom, and the United States—proposed a "Concept for a Multilateral Mechanism for Reliable Access to Nuclear Fuel"[66] (referred to here as the Six Country Concept). This proposal reportedly developed from a U.S. initiative following President Bush's 2004 proposal. It would not require states to forgo enrichment and reprocessing, but participation would be limited to those states that did not currently have enrichment and reprocessing capabilities.

The Six Country Concept calls for a multi-tiered backup mechanism to ensure the supply of low enriched uranium (LEU) for nuclear fuel. The proposal would work as follows: (1) A commercial supply relationship is interrupted for reasons other than nonproliferation; (2) The recipient or supplier state can approach the IAEA to request backup supply; (3) The IAEA would rule out commercial or technical reasons for interruption (to avoid a market disruption) and assess whether the recipient meets the following qualifications: it must have a comprehensive safeguards agreement and Additional Protocol in force; it must adhere to international nuclear safety and physical protection standards; and it is not pursuing sensitive fuel cycle activities (which are not defined); (4) The IAEA would facilitate new arrangements with alternative suppliers.

Two mechanisms were proposed to create multiple tiers of assurances: including a standard backup provision in commercial contracts, and establishing reserves of LEU (not necessarily held by the IAEA, but possibly with rights regarding the use of the reserves). The Six Country Concept addressed several future options, all of which are longer-term in nature. They include providing reliable access to existing reprocessing capabilities for spent fuel management; multilateral cooperation in fresh fuel fabrication and spent fuel management; international enrichment centers; and new fuel cycle technology development that could incorporate fuel supply assurances.

World Nuclear Association

In May 2006, the private-sector World Nuclear Association (WNA) Working Group on Security of the International Nuclear Fuel Cycle outlined proposals for assuring front-end and back-end nuclear fuel supplies.[67] Like the Six Country Concept, the WNA proposal envisions a system of supply assurances that starts first with normal market procedures attempting to reestablish nuclear fuel supply after interruptions. Also similar to the Six Country proposal, a pre-established network of suppliers could be triggered through the IAEA if supply were interrupted for political reasons. If that network then failed, stocks held by national governments could be used.

The first tier of assurances, therefore, is through commercial suppliers. The second level of supply commitment would use a "standard backup supply clause" in enrichment contracts, supported by governments and the IAEA. "To ensure that no single enricher is unfairly burdened with the responsibility of providing backup supply, the other (remaining) enrichers would then supply the contracted enrichment in equal shares under terms agreed between the IAEA and the enrichers," according to the proposal.

For fuel fabrication, a backup supply system would be more complicated, according to the WNA report. "Because fuel design is specific to each reactor design, an effective mechanism would require stockpiling of different fuel types/designs. The cost of such a mechanism could thus be substantial," according to the report. However, WNA noted that unlike uranium enrichment technology, uranium fuel fabrication is not of proliferation concern.

The WNA report also noted the need for back-end nuclear fuel cycle supply assurances, to prevent a future scenario in which reprocessing technologies spread as nuclear power programs expand. The report recommends that a clear option to reprocess spent fuel at affordable prices be offered to states that do not have indigenous reprocessing programs. Such assurances would be part of a longer-term approach.

Japan's IAEA Standby Arrangements System

Japan presented a "complementary proposal" to the Six Country Concept at the IAEA in September 2006. Japan's concerns with the Six Country Concept centered on the implication that it would deny the right for states to use nuclear technology for commercial purposes and because it assured the supply only of LEU, rather than all front-end nuclear fuel cycle services. Japan proposed instead to create an "IAEA Standby Arrangements System" that would act as an early warning system to prevent a break in supply to recipients. With a list of supply capacities from each state updated annually and a virtual bank of front-end fuel cycle services (from natural uranium to fuel fabrication), the IAEA would facilitate supply to recipient states before supply was completely stopped. States determined by the IAEA Board of Governors to be in good nonproliferation standing by the IAEA could participate.

UK "Nuclear Fuel Assurance (NFA)"

The United Kingdom has proposed a political assurance of non-interference in the delivery of commercial nuclear contracts, called a nuclear fuel assurance (NFA). This concept incorporates an earlier proposal that would creates "enrichment bonds" to give advance assurance of export approvals for nuclear fuel to recipient states. As the UK Prime Minister's report, *Road to 2010,* summarized the NFA: "The UK's Nuclear Fuel Assurance is complementary to other proposals put forward and provides a guarantee that export licences for nuclear fuel enrichment services would only be withheld in the event of non-compliance with non-proliferation obligations."[68]

The NFA would be a government-to-government agreement between supplier state or states and the recipient state, with the IAEA as co-signatory. The supplier government would guarantee that, subject to the IAEA's determination that the recipient was in good nonproliferation standing, national enrichment providers will be given the necessary export approvals to supply the recipient states. It is a transparent legal mechanism designed to give further credible assurance of supply with a "prior consent to export" arrangement. The IAEA would make the final decision on whether conditions had been met to allow the export of LEU.[69] A model agreement to be used as a standard for an NFA could be adopted by the Board of Governors. The NFA concept has been presented to the IAEA Board of Governors and is still under discussion

Assurance of Fuel Supply: Fuel Reserves

A fuel reserve is meant to appeal to countries concerned about a possible cut-off in their nuclear fuel supply for reasons unrelated to nonproliferation, such as a non-commercial or political dispute with the supplier country. A fuel bank under IAEA control has achieved initial funding pledges but is still under consideration by the IAEA Board of Governors. The IAEA Board approved terms for a Russian-operated "fuel reserve" in November 2009. Additionally, the United States has declared that it would downblend excess military HEU to LEU and hold it in reserve as part of the Six-Country Concept, under the Department of Energy's Reliable Fuel Supply program. These three proposals are discussed below.

IAEA LEU Fuel Bank

In September 2006, former Senator Sam Nunn, co-chairman of the Nuclear Threat Initiative (NTI),[70] announced NTI's pledge of $50 million as seed money to create a low-enriched uranium stockpile owned and managed by the IAEA. NTI believes that the establishment of such an LEU reserve would assure an international supply of nuclear fuel on a non-discriminatory, nonpolitical basis to recipient states.

Provision of the NTI money was contingent on the IAEA taking the necessary preparatory actions to establish the reserve and on contribution of an additional $100 million or an equivalent value of LEU by one or more IAEA Member States.[71] The latter condition was met in March 2009. The U.S. Congress approved $50 million for an international fuel bank in December 2007 (see below). Norway pledged $5 million to the fuel bank in February 2008.[72] The United Arab Emirates announced a contribution of $10 million on August 1, 2008.[73] The European Union pledged 25 million euros in December 2008,[74] and Kuwait pledged $10 million in March 2009.[75] To date, the IAEA has received the U.S. government contribution, which is being held in a suspense account until the Board of Governors approves the LEU bank.

No policy conditions have been set by donor countries—policy questions are meant to be solved by the IAEA and member states. Key issues still to be determined include the reserve's content, location, criteria for determining access to the stocks including safety and export control standards, the fuel's pricing, and how the fuel in reserve would be fabricated into the appropriate fuel type for the customer's reactor.[76] Kazakhstan proposed in May 2009 that it host the fuel bank.

The IAEA secretariat drew up draft plans for the fuel bank, which were presented to the Board of Governors at its June 2009 meeting along with a proposal for a Russian-hosted bank and the German-proposed multilateral enrichment project (see below). However, developing countries reportedly rejected the director general's proposal to negotiate details and approve these arrangements at the September Board meeting. Opponents and skeptics cite concerns that their legal rights under the NPT to develop fuel cycle facilities would be infringed upon if such facilities were established. Proponents counter that these arrangements would be optional, but are meant to give countries alternatives to developing their own fuel cycle capabilities.[77] As of early 2010, the IAEA secretariat was engaged in discussing questions and concerns about the fuel bank proposal with member states.

The House (H.R. 1585) and Senate versions (S. 1547) of the National Defense Authorization Act for Fiscal Year 2008 both authorized $50 million to be appropriated to the

Department of Energy for the "International Atomic Energy Agency Nuclear Fuel Bank." The conference report supports the establishment of a fuel bank and notes that "additional work will be required in order to provide appropriate guidance to the executive branch regarding criteria for access by foreign countries to any fuel bank established at the IAEA with materials or funds provided by the United States."[78]

Both the House (H.R. 2641) and Senate (S. 1751) Energy and Water Appropriations bills for FY2008 recommended funding for an international nuclear fuel bank under the IAEA, and proposed making available $100 million and $50 million respectively. The Consolidated Appropriations Act for FY2008, which became P.L. 110-161 on December 26, 2007, provided that $50 million should be available until expended for "the contribution of the United States to create a low-enriched uranium stockpile for an International Nuclear Fuel Bank supply of nuclear fuel for peaceful means under the International Atomic Energy Agency." On August 4, 2008, the U.S. Secretary of Energy issued an official letter to the IAEA donating "nearly 50 million" to the international nuclear fuel bank.[79] This reflects a congressionally mandated rescission that was applied proportionally across the Department of Energy's budget.[80]

Russian LEU Fuel Reserve, Angarsk

The IAEA Board of Governors in November 2009 authorized the director general to sign an agreement with Russia establishing an LEU fuel reserve at Angarsk (GOV/2009/81). The fuel is to be available to a country facing a disruption of supply "unrelated to technical or commercial reasons."[81] The reserve would contain about 120 MT of LEU in the form of UF_6 with an enrichment level ranging from 2 to 4.95% and would be under IAEA safeguards.[82] The Russian government would cover the cost of safeguards and all operating costs.

The Russian plan envisions that countries facing a fuel supply cut-off would apply to the IAEA to access the fuel reserve. The director general would assess whether the country meets the criteria for access. To qualify, the potential recipient state would have to be a non-nuclear-weapon state member of the IAEA with a safeguards agreement in effect. To release the material, the IAEA would have to confirm that all nuclear material was accounted for in the state, there was no indication of diversion of material, and no safeguards issues were under review by the Board of Governors. If the criteria were met, the director general would ask Russia to release fuel to that country. The recipient country would pay market rates for the uranium. A country would not have to waive its right to develop its own fuel cycle capabilities in order to access the fuel reserve.

U.S. LEU Fuel Reserve: "Reliable Fuel Supply" Program

In 2005, then-Secretary of Energy Samuel Bodman announced that 17.4 metric tons of U.S. surplus highly enriched uranium would be downblended to low-enriched uranium to be used as a U.S. fuel reserve. The goal of the U.S. reserve is to supply fuel in the event of a disruption unrelated to proliferation. Secretary Bodman described the U.S. reserve as supporting the "twin goals expanding the use of nuclear power and curbing nuclear proliferation," and said its purpose was to "help countries to pursue nuclear power confidently, without the burden of producing their own fuel, while curbing the spread of sensitive technology."[83]

The precise terms and conditions governing the release of the fuel and its potential recipients have not yet been determined. According to U.S. officials, the material designated for the U.S. reserve would be kept under national control, and not be part of the IAEA fuel bank.[84] Stringent U.S. requirements on U.S.-origin material, pursuant to the 1954 Atomic Energy Act (as amended), include safeguards in perpetuity, prior consent for enrichment and reprocessing, and the right of return should a non-nuclear-weapon state detonate a nuclear explosive device.[85]

The DOE's Fissile Material Disposition program manages this initiative, called "Reliable Fuel Supply." WesDyne International, LLC, and Nuclear Fuel Services, Inc., were awarded a contract in 2007 to downblend and store the material. Nuclear Fuel Services began downblending in December 2009 at its facility in Erwin, TN, and is expected to complete the work in 2010 or early 2011. The 17.4 MT of HEU is expected to produce about 290 MT of low enriched uranium. According to an NNSA press release, WesDyne will sell a "small fraction" of the resulting low enriched uranium on the market over a three to four year period to cover the project's costs.[86] According to the FY2011 Department of Energy budget request, in 2009 the program completed all shipments of HEU for the Reliable Fuel Supply Initiative, and enough LEU has been downblended to date for multiple commercial reactor core reloads.

In addition to the downblending of 17.4 MT of HEU, an additional 12.1 MT is to be downblended to approximately 220 MT of LEU "to provide assurance of fuel supply to utilities participating in the MOX program for the disposition of surplus weapons plutonium."[87] This tranche is expected to be downblended by 2012.

LEU from both of the above HEU disposition programs will be stored until needed at Westinghouse's Columbia Fuel Fabrication Facility in Columbia, SC.

Assurance of Supply: Enrichment Services

Proposals to assure supply and prevent the spread of enrichment technology have also included the creation of commercially based multinational uranium enrichment centers. The Russian Federation has made progress in establishing an International Uranium Enrichment Center (IUEC) at Angarsk. Germany has proposed an internationally owned and operated Multilateral Enrichment Sanctuary Program. Some point out that URENCO, a joint German, Dutch and British consortium, has demonstrated a multilateral commercial model for uranium enrichment since the 1970s. Some countries are concerned that giving support for multilateral-owned facilities would undermine their rights to nuclear technology for peaceful purposes under the NPT, and view the only solution to energy security as being an independent fuel cycle. However, as this may not be economically viable for most countries, multilateral solutions continue to be attractive.

International Uranium Enrichment Center (IUEC), Angarsk, Russia

Russia has created a model International Uranium Enrichment Center (IUEC) at Angarsk (approximately 3,000 miles east of Moscow).[88] The Angarsk IUEC began operation on September 5, 2007. Kazakhstan was the first partner,[89] Armenia and Ukraine have also signed intergovernmental agreements to join IUEC. Finland, South Korea, and Belgium are negotiating their memberships in the IUEC.[90] As part of an open joint-stock company, IUEC participants would receive dividends from IUEC profits.

To join the Angarsk IUEC, countries must agree that the material be used for "nuclear energy production."[91] The IUEC is "chiefly oriented to States not developing uranium enrichment capabilities on their territory."[92] Russia now includes the IUEC on its list of Russian facilities that could be placed under IAEA safeguards. Additional documents for implementing safeguards at the facility are now under discussion. The safeguards would apply to the nuclear material at the facility.[93]

Germany's Multilateral Enrichment Sanctuary Project (MESP)

Germany proposed in May 2007 that a new enrichment facility be built and placed under IAEA ownership in an extraterritorial area.[94] An independent management board or consortium would finance and run the plant on a commercial basis, but the IAEA would decide whether to supply enriched fuel according to nonproliferation criteria. Germany argues that this approach is advantageous since it does not prohibit uranium enrichment, but does provide a commercially viable, politically neutral option for fuel supply and could create competition on the world market by creating a new fuel service provider. With an economically viable option on neutral ground, it will be harder for states to justify starting their own enrichment program for commercial reasons. A proposal was presented to the IAEA Board of Governors in June 2009, and a draft MESP agreement (INFCIRC/765) submitted to the IAEA in July 2009. This concept has not yet been approved by the Board of Governors but is under active discussion.

Back-End Fuel Cycle Proposals

Multilateral proposals for the end of the fuel cycle are less developed at this stage. On-site storage of spent fuel is most common, and some countries reprocess their spent fuel rods into mixed-oxide fuel. Multilateral solutions to the back-end issues are also motivated by the idea of preventing the further spread of reprocessing technology, which could be used for the separation of plutonium for weapons purposes. As with the debate over uranium enrichment technology, while reprocessing technology may be too expensive or technologically out of reach for many states at this stage, countries are hesitant to agree to multilateral approaches that may be interpreted as giving up their right to develop this technology for peaceful purposes. There are no multilateral reprocessing facilities now proposed.

Another proposal has been the establishment of an international spent fuel repository, perhaps in Russia. While Russian law allows for the import of waste, the government of Russia has not yet proposed such a facility, partly due to potential public opposition.

States also cooperate on joint research ventures on advanced and fast reactors such as the Generation IV International Forum (GIF) or IAEA's INPRO. A major U.S.-led initiative, the Global Nuclear Energy Partnership (GNEP) was meant to inspire international collaboration on developing a proliferation-resistance closed fuel cycle, discussed below.

Global Nuclear Energy Partnership (GNEP)

In February 2006, U.S. Secretary of Energy Bodman announced the Global Nuclear Energy Partnership (GNEP), drawing together two of the former Bush Administration's policy goals: promotion of nuclear energy and nonproliferation. Recycling nuclear fuel to produce more energy and reduce waste and encouraging global prosperity were a few of DOE's stated aims for the program. GNEP built on DOE's Advanced Fuel Cycle Initiative (AFCI), a program that began in 2003 to develop and demonstrate spent fuel reprocessing/recycling technology. Today, GNEP continues in its international capacity, but funding for the domestic portions of the program are now only provided for fundamental research and development work. Efforts under AFCI to develop large-scale reprocessing and recycling facilities have been halted by the Obama Administration, and the program has been renamed Fuel Cycle Research and Development.

The domestic component of GNEP originally focused on the future of nuclear energy in the United States: What kind of future reactors will be licensed, and how spent nuclear power reactor fuel will be handled. Existing commercial light water reactors were expected to continue as the predominant technology for at least the next two decades. Spent fuel from existing reactors was to be stored or retrievably emplaced at the planned Yucca Mountain, NV, repository, awaiting possible future reprocessing and recycling.

Reprocessing facilities were to use new technologies developed by AFCI to avoid separation of pure plutonium that could be used for weapons. However, there has been continuing controversy over how proliferation-resistant such processes might be. High-level waste from reprocessing (mostly fission products) would have gone to the Yucca Mountain repository, and the recycled plutonium and uranium was to be fabricated into fuel for an Advanced Burner Reactor, a fast reactor to be developed by DOE's Generation IV Nuclear Energy Systems Initiative. In the longer term, plutonium and other transuranics (elements heavier than uranium) in spent fuel were to be fabricated into new fuel for future fast reactors. Eventually, that fuel was to be continually recycled until all the transuranics were consumed, leaving the fission products to be disposed of in a geologic repository.

The international component of GNEP as originally envisioned was a consortium of nations with advanced nuclear technology that would provide fuel services and reactors to countries that "refrain" from fuel cycle activities, such as enrichment and reprocessing. It was essentially a fuel leasing approach, wherein the supplier would take responsibility for the final disposition of the spent fuel. This could mean taking back the spent fuel, but might also mean, according to DOE, that the supplier "would retain the responsibility to ensure that the material is secured, safeguarded and disposed of in a manner that meets shared nonproliferation policies."[95] While this describes the responsibility of the supplier, the vagueness of the language suggests that any number of solutions, including on-site storage, could be the outcome.

GNEP foresaw a system whereby supplier states would take back spent fuel, but many nations lack the political will to do so. Skeptics raised the question of whether the technology used in GNEP would have been a net gain for nonproliferation efforts, since the United States does not reprocess or re-use plutonium now. In their view, the "proliferation-resistance" of technologies under consideration must be assessed against the major alternative: disposal of sealed, intact fuel rods in a geologic repository. At the direction of the White House, Energy Secretary Steven Chu established the Blue Ribbon Commission on America's Nuclear Future on January 29, 2010, to recommend a new national strategy for managing spent nuclear fuel and high-level waste, including an examination of reprocessing and recycling options.

Much of the AFCI's research had focused on a separations technology called UREX+, in which uranium and other elements are chemically removed from dissolved spent fuel, leaving a mixture of plutonium and other highly radioactive elements. Proponents believe UREX+ is proliferation-resistant, because further purification would be required to make the plutonium useable for weapons and because its high radioactivity would make it difficult to divert or work with. In contrast, conventional reprocessing using the PUREX process can produce weapons-useable plutonium that can be processed in unshielded gloveboxes.

However, critics see the potential nonproliferation benefits of UREX+ over PUREX as minimal. Richard Garwin suggested in testimony to Congress in 2006 that Urex+ fuel fails the proliferation-resistance test. Since it contains 90% plutonium, it could be far more attractive to divert than current spent fuel, which contains 1% plutonium. In other words, a terrorist would only have to reprocess 11 kg of Urex+ fuel to obtain roughly 10 kg of plutonium, in contrast to reprocessing 1,000 kg of highly radioactive spent fuel to get the same amount from light water reactor fuel.[96] Under the Obama Administration, the Fuel Cycle Research and Development program is no longer focusing on Urex processes but is instead conducting long-term research that could support a broad range of technology options, according to the DOE FY2011 budget justification.

Another nonproliferation-related concern about GNEP was how its implementation would have affected global stockpiles of separated plutonium. Frank Von Hippel pointed to costly plutonium recycling programs in the United Kingdom., Russia, and Japan, where separated plutonium stocks have accumulated to 250 tons, enough for 30,000 nuclear warheads. In Von Hippel's view, GNEP would have exchanged the safer on-site spent fuel storage at reactors for central storage of separated transuranics and high-level waste, cost many times more, and increased the global plutonium stockpile.[97]

A separate set of questions focused on how effective GNEP would have been in achieving its goals. By offering incentives for the back end of the fuel cycle, GNEP was designed to attract states to participate in the fuel supply assurances part of the framework. However, back-end fuel cycle assurances would require significant changes in policies and laws, as well as efforts to commercialize technologies. Further, it is far from clear that all suppliers would be able to offer the full range of fuel cycle assurances, raising the question of the relative competitiveness of suppliers. Critics did not necessarily argue that the overall vision of GNEP was misplaced, but were generally skeptical that its vision could be achieved, particularly in the timeframe proposed.

GNEP itself marked a departure from a U.S. policy of not encouraging the use of plutonium in civil nuclear fuel cycles. Supporters suggested that the U.S. policy developed in the late 1970s did not envision a recycling process that would not separate pure plutonium, and therefore questioned the underlying assumptions of that longstanding policy. Critics of GNEP have suggested that even though many nations did not agree with the United States in the 1970s on the dangers of having stockpiles of separated plutonium, the message that the United States conveyed was that reprocessing was unnecessary to reap the benefits of nuclear power and that GNEP conveyed the opposite message. Moreover, some critics pointed to the accumulation since the 1970s of separated plutonium as a particular threat, given the potential for terrorist interest in acquiring nuclear material.

An October 2007 study published by the National Academies of Science recommended that research and development activities for new reprocessing plants continue but be scaled back, with more time and peer review before commercial plants are built. It strongly criticized DOE's timeline for the program, saying that "achieving GNEP's goals are too early in development to justify DOE's accelerated schedule for construction of commercial facilities that would use these technologies."[98] Congress also expressed significant concerns about GNEP in recent years, particularly over the Bush Administration's ambitious schedule for developing fuel cycle demonstration facilities by FY2020.

The GNEP proposal attracted some international interest, at least among potential supplier states. Officials from China, France, Japan, Russia, and the United States met in Washington, DC, on May 21, 2007, to discuss GNEP and its goals. According to a joint statement issued after the meeting, "The participants believe in order to implement the GNEP without prejudice to other corresponding initiatives, a number of near- and long-term technical challenges must be met. They include development of advanced, more proliferation resistant fuel cycle approaches and reactor technologies that will preserve existing international market regulations."

In a formal presentation of GNEP principles, made September 16, 2007, in Vienna, Austria, participation was opened to all nations on a voluntary basis that agreed to internationally accepted standards for a safe, peaceful, and secure nuclear fuel cycle.[99] Sixteen countries joined the United States in signing the Statement of Principles for GNEP in September 2007, and the organization's website listed 25 "GNEP Partners" in June 2009.[100] The principles call for safe expansion of nuclear energy, enhanced nuclear safeguards, international supply frameworks, and development of fast reactors, "more proliferation resistant" nuclear power reactors, and spent fuel recycling technologies in facilities that do not separate pure plutonium. They did not call upon states to renounce or refrain from indigenous development of enrichment or reprocessing technologies but emphasized the goal of creating "a viable alternative to acquisition of sensitive fuel cycle technologies." It further emphasized that participants would not be giving up any rights to benefit from peaceful nuclear energy.

It may be difficult for the United States and others to define which states are suppliers and which are recipients. Informally, U.S. policy currently recognizes 10 states as having enrichment capability—the five nuclear weapon states (United States, United Kingdom, France, China, Russia) plus Japan, Argentina, Brazil, the Netherlands, and Germany. While Argentina has a plant (Pilcaniyeu) under safeguards, this plant has never operated commercially and it is doubtful that it will be cost-effective, since it uses outdated gaseous diffusion technology. Brazil's centrifuge enrichment plant at Resende is still in the early stages of commissioning and won't produce at a commercial scale for several years. States such as Australia, Canada, South Africa, and Ukraine have stated they would be interested in developing enrichment capability for export. On the reprocessing side, South Korea has expressed interest in becoming a GNEP supplier state through development of a pyroprocessing technique that does not separate plutonium from uranium. In the past, the United States for proliferation reasons has rejected requests from South Korea to reprocess U.S.-origin spent fuel.

Despite the change in U.S. administrations, the international GNEP members have continued to meet. The fourth Global Nuclear Energy Steering Group meeting was held in Tokyo, Japan on April 7-8, 2009. GNEP's Infrastructure Development Working Group

(IDWG) met during May 18-20, 2009, in Manchester, England, with 18 GNEP partner and observer countries and representatives from observer organizations participating, according to the GNEP website. The third GNEP Executive Committee Ministerial-Level Meeting was held in Beijing on October 23, 2009, and the IDWG met in Vienna, Austria, on December 9-10, 2009.

Supply-Side Approaches

Discussions in the Nuclear Suppliers Group (NSG)

Current NSG Guidelines say supplier countries should exercise restraint in transferring any enrichment or reprocessing technologies. In the past several years, negotiations have been under way to specify criteria for such transfers. Since the 1970s, NSG members have adhered to an informal restriction on transferring enrichment, reprocessing, and heavy water technology to states outside the NSG. France first proposed an approach that would lay out a set of criteria that recipient states would first need to meet, including the following:

- Member of the NPT in full compliance.
- Comprehensive safeguards agreement and Additional Protocol in force.
- No breach of safeguards obligations, no IAEA Board of Governors decisions taken to address lack of confidence over peaceful intentions.
- Adherence to NSG Guidelines.
- Bilateral agreement with the supplier that includes assurances on non-explosive uses, effective safeguards in perpetuity, and retransfer.
- Commitment to apply international standards of physical protection.
- Commitment to IAEA safety standards.

The NSG also discussed including more subjective criteria in a decision to supply a state with fuel cycle technology such as general conditions of stability and security, potential negative impact on the stability and security of the recipient state, and whether there is a credible and coherent rationale for pursuing enrichment and reprocessing capability for civil nuclear power purposes.

No consensus has yet been reached on how to define these criteria, and a number of questions remain. For example, it is clear from these requirements that states outside the NPT—such as India, Pakistan, and Israel—would be prohibited from reprocessing and enrichment cooperation with NSG members. To add further complications, the nuclear cooperation agreement (so-called 123 Agreement) between the US and India provides consent in principle for India to reprocess U.S. spent fuel and agreement in principle to transfer enrichment and reprocessing-related technology to India, pursuant to an amendment to the agreement.[101] These two details suggest that India is a reprocessing technology holder, despite not having its reprocessing facilities under comprehensive IAEA safeguards, and call into question criteria for distinguishing between states that should receive assistance and those that should not, particularly since India is neither a party to the NPT nor an NSG member.

The United States had objected to the criteria-based approach, favoring a moratorium, until spring 2008 when the Bush administration changed its policy. The U.S. reportedly would be in favor of establishing criteria if they included a requirement that uranium enrichment be exported only through a "black box" arrangement.[102] Canada objects to this, and negotiations are still underway.[103]

Group of Eight Nations (G-8)

The Group of Eight (G-8) Nations has been the forum where joint policy statements have been made on this issue in recent years. From 2004-2007, the Group of Eight (G-8) Nations announced a year-long suspension of any such transfers at their annual summit meetings. The 2008 Summit declaration stated:

> We agree that transfers of enrichment equipment, facilities and technology to any additional state in the next year will be subject to conditions that, at a minimum, do not permit or enable replication of the facilities; and where technically feasible reprocessing transfers to any additional state will be subject to those same conditions.[104]

Since in 2009 there still was no agreement on the transfer criteria in the NSG, the G-8 countries said that they would implement this policy on a national basis:

> To reduce the proliferation risks associated with the spread of enrichment and reprocessing facilities, equipment and technology, we welcome the progress that continues to be made by the Nuclear Suppliers Group (NSG) on mechanisms to strengthen controls on transfers of such enrichment and reprocessing items and technology. While noting that the NSG has not yet reached consensus on this issue, we agree that the NSG discussions have yielded useful and constructive proposals contained in the NSG's "clean text" developed at the 20 November 2008 Consultative Group meeting. Pending completion of work in the NSG, we agree to implement this text on a national basis in the next year. We urge the NSG to accelerate its work and swiftly reach consensus this year to allow for global implementation of a strengthened mechanism on transfers of enrichment and reprocessing facilities, equipment, and technology.[105]

COMPARISON OF PROPOSALS

Table 4 provides a comparison of the major proposals currently in circulation to restrict sensitive nuclear fuel technology development. The table is based on one created by Chaim Braun presented at the September 2006 IAEA conference on nuclear fuel supply assurances.[106]

Table 4. Comparison of Major Proposals on Nuclear Fuel Services and Supply Assurances

	IAEA/INFCIRC/640	Putin Initiative	GNEP	Six Country Concept	World Nuclear Association
Goals	Identify multilateral approaches across the fuel cycle; improve non-proliferation assurances without disrupting market mechanisms.	Establish international commercially operated nuclear fuel service centers in Russia, to include enrichment, education and training, and spent fuel management.	Enable expansion of nuclear power in the United States and around the world, promote nuclear nonpro-liferation goals, and help resolve nuclear waste disposal issues. Provide states with front-end and back-end services, to provide an alternative to the creation of national enrichment and reprocessing capabilities.	Create interim measures for front-end assurances.	Enhance supply security.
Target	Front-end and back-end services including uranium enrichment, fuel reprocessing, and disposal and storage of spent fuel.[a]	Supply of nuclear fuel and possibly other fuel cycle services.	Front- and back-end services. It could create new class of "reactor-only" states.	Supply of nuclear fuel.	Primarily fuel supply.
Methods	Reinforce commercial contracts with transparent supplier arrangements with government backing. International supply guarantees backed by fuel reserves.	Commercial, long-term contracts; recipients will have limited control over joint ventures. IAEA will be involved.[b]	Use existing enrichment and reprocessing services; develop more prolife-ration-resistant technology. Fuel supplier will be responsible for spent fuel disposition.	Level I: Market Level II: Fuel assurance mechanism at IAEA Level III: Mutual commercial back-up arrangements Level IV: Enriched uranium reserves	Level I: Market meets demand Level II: Standard back-up supply clause in enrichment contracts, with IAEA assurances Level III: Gov't stocks of enriched uranium
IAEA Role	IAEA participates in administering supply guarantees, possibly as	IAEA would ensure supply with fuel bank created by purchasing existing fuel stocks	IAEA would apply safeguards.	IAEA as broker. IAEA assesses status of safeguards agreements,	IAEA would approve "triggering"

Table 4. (Continued)

	IAEA/INFCIRC/640	Putin Initiative	GNEP	Six Country Concept	World Nuclear Association
	guarantor of service supplies with use of a fuel bank. Possible IAEA supervision of an interna-tional consortium for reprocessing services.	and placing them under its control (IAEA would receive new funding to do so).		safeguards Implemen-tation, safety, physical protection and whether a country is pursuing sensitive fuel cycle activities.	mechanism for supply back-up. IAEA could manage enriched uranium reserve.
Eligibility	Recipient countries would renounce the construction and operation of sensitive fuel cycle facilities and accept safeguards of the highest current standards including comprehensive safeguards and the Additional Protocol.	Equal access, but prere-quisite is compliance with the nonproliferation regime. Potential provider states could include Australia and Canada.	No requirements now (versus initial requirement for recipient states to forego enrichment and reprocessing).	IAEA-approved states that are in good NPT standing. States that develop national capabilities will not be eligible.	IAEA-approved states that meet all NPT obligations.
Role of Industry	Managing, operating centers.	Performing fuel services at designated center.	Performing fuel services, but not necessarily coordinated.	Perform enrichment contracts; identified need to address back-end of fuel cycle.	Perform enrichment contracts. No new capabilities required.
Potential Concerns	No mechanism specified for assessing state's nonproliferation record.	Incentives not specified, as well as compliance with nonproliferation regime. Unclear how commitments to forgo sensitive fuel cycle activities will be incorporated into contracts.	Lack of political will to take back spent fuel. Concerns about gains for Nonpro-liferation, if the United States was not reprocessing to begin with.	Incentives may be insufficient.	Incentives may be insufficient. How to determine price on enriched uranium reserves, if they are required.

a. INFCIRC/640, p. 103.

b. "Questions Abound on Proposals by Bush, Putin on Fuel Centers," *Nuclear Fuel*, March 13, 2006, vol. 31, no. 3.

PROSPECTS FOR IMPLEMENTING FUEL ASSURANCE MECHANISMS

Proposals to provide an international and institutional framework for peaceful nuclear activities have abounded since the 1940s, but few have been implemented. The U.S.-sponsored Baruch Plan introduced at the United Nations in 1946 recommended establishing an international agency with managerial control or ownership of all atomic energy activities. The International Atomic Energy Agency, established in 1957, emerged as a paler version of what was suggested in the Baruch Plan, but still retains authorities in its statute to store fissile material.

Concern about proliferation led to a flurry of proposals in the 1970s and 1980s as the United States and others convened groups to study the issues.[107] One idea studied in the mid-1970s was regional nuclear fuel cycle centers, focused on reprocessing technologies. Several factors contributed to its lack of success, despite support by the U.S. Congress: low uranium prices (making plutonium recovery relatively unattractive), a slump in the nuclear industry in the late 1970s and early 1980s, and U.S. opposition to reprocessing from the late 1970s. Member states of the IAEA also convened the International Fuel Cycle Evaluation project, which involved 60 countries and international organizations. INFCE working group reports suggested establishing a multi-tiered assurance of supply mechanism similar to the one proposed by the Six Country Concept in 2006. States also studied international plutonium storage in the late 1970s and early 1980s, but could not agree on how to define excess material or the requirements for releasing materials.

As in the past, the success of current proposals may depend on whether nuclear energy is truly revived not just in the United States, but globally. That revival will likely depend on significant support for nuclear energy in the form of policy, price supports, and incentives. Factors that may help improve the position of nuclear energy against alternative sources of electricity include higher prices for other sources (natural gas and coal through a carbon tax or other restrictions), improved reactor designs to reduce capital costs, regulatory improvements, and waste disposal solutions.

The willingness of fuel recipient states to participate in international enrichment centers rather than develop indigenous enrichment capabilities, and confidence in fuel supply assurance mechanisms such as an international fuel bank, will largely determine the success of the overall policy goal—to prevent further spread of enrichment and reprocessing technologies. So far, proposals addressing this challenge have originated in the supplier states, with many recipient states continuing to voice concern that their right to peaceful nuclear energy technology under the NPT is in jeopardy. Increasingly, however, participation is being presented as a market-based decision by countries to refrain, at least for the present, from developing their own fuel enrichment programs.

Another factor that will shape the success of these proposals is the possible addition of other incentives. Simply making nuclear energy cost-effective may not induce countries to forgo indigenous enrichment and reprocessing. Such decisions may require other incentives, perhaps even outside the nuclear realm, to make them palatable. The experience of Iran may be instructive here. Russia's offer to provide assured enrichment services on Russian soil has gone nowhere; instead, other, broader trade incentives may be necessary. While the case of Iran may illustrate the extreme end of the spectrum, in terms of a country determined to develop a capability for a weapons program, non-nuclear-weapon states will clearly take notice of how a solution develops for Iran.

ISSUES FOR CONGRESS

Congress would have a considerable role in at least four areas of oversight related to fuel cycle proposals. The first is providing funding and oversight of U.S. domestic programs related to expanding nuclear energy in the United States. Key among these programs are nuclear research and development programs and federal incentives for building new commercial reactors.[108]

The second area is policy direction and/or funding for international measures to assure supply. What guarantees should the United States insist upon in exchange for helping provide fuel assurances? Although the Six Country Concept contains an option for a fuel bank, it would not require participants to forswear enrichment and reprocessing.

A third set of policy issues may arise in the context of implementing the international component of GNEP. As referenced above, in the original policy documents, GNEP participant states would "agree to refrain from fuel cycle initiatives." However, in recent ministerial meetings, this language was no longer used and participation was opened to all. This is most likely meant to increase participation in the initiative by emphasizing that GNEP is not asking states to give up rights to peaceful nuclear technology.

Some observers believe that further restrictions on non-nuclear-weapon states party to the NPT are untenable in the absence of substantial disarmament commitments by nuclear weapon states. In particular, a January 4, 2007, *Wall Street Journal* op-ed by George Shultz, Bill Perry, Henry Kissinger, and Sam Nunn, entitled "A World Free of Nuclear Weapons," noted that non-nuclear-weapon states have grown increasingly skeptical of the sincerity of nuclear weapon states in this regard. Some observers have asserted that non-nuclear-weapon states will not tolerate limits on NPT Article IV rights (right to pursue peaceful uses of nuclear energy) without progress under Article VI of the NPT (disarmament). Amending the NPT is seen by most observers as unattainable. President Obama called for the eventual elimination of nuclear weapons in a speech in the Czech Republic on April 5, 2009.

The IAEA experts group report, INFCIRC/640, did point to the political usefulness of achieving a ban on producing fissile material for nuclear weapons (known as fissile material production cutoff treaty, or FMCT) to provide more balance between the obligations of nuclear and nonnuclear-weapon states. Obama administration officials have indicated that they will pursue negotiations on a fissile material cut-off treaty that includes verification provisions. Ultimately, any such treaty would require Senate advice and consent to ratification.

A fourth area in which Congress plays a key role would be in the approval of nuclear cooperation agreements.[109] In some cases, the United States may seek additional reassurances regarding fuel cycle facilities during negotiations of civilian nuclear cooperation agreements. This was a topic of controversy during the approval process for the civilian nuclear cooperation agreement with India in September 2008.[110] A civilian cooperation agreement with the United Arab Emirates was submitted to the 111th Congress for consideration on May 21, 2009,[111] and it took effect December 17, 2009. In April 2009, the UAE had signed a memorandum of understanding with the United States saying it would forgo "domestic enrichment and reprocessing capabilities in favor of long-term commitments of the secure external supply of nuclear fuel." In addition, the nuclear cooperation agreement's text itself states that the United States can end nuclear cooperation with the UAE if it acquires enrichment or reprocessing facilities.

Acknowledgments

Jill Marie Parillo and Sharon Squassoni were original contributors to this chapter.

End Notes

[1] U.S. Department of Energy, "The Global Nuclear Energy Partnership," Factsheet 06-GA50506-01

[2] See also CRS Report RL34248, *Pakistan's Nuclear Weapons: Proliferation and Security Issues*, by Paul K. Kerr and Mary Beth Nikitin.

[3] "Security Council, in Presidential Statement, Underlines Importance of Iran's Re-Establishing Full, Sustained Suspension of Uranium Enrichment Activities," March 29, 2006, at http://www.un.org/News/Press/docs/2006/ sc8679.doc.htm, and U.N. Security Council Resolution 1737 (2006) http://www.un.org/News/Press/ docs//2006/ sc8928.doc.htm.

[4] Most observers point to the obligation in Article IV that such pursuit must be consistent with a state's obligations under Articles II and II of the treaty. Article II refers to a state's obligation to foreswear nuclear weapons development and Article III refers to a state's obligation to undertake safeguards "for the exclusive purpose of verification of the fulfillment of its obligations" under the treaty.

[5] See timeline of fuel cycle proposals, available at http://www.iaea.org/News Center/Focus/ FuelCycle/ key_events.shtml.

[6] This section was prepared by Mark Holt, Specialist in Energy Policy, and Anthony Andrews, Specialist in Energy Policy, in the Resources, Science, and Industry Division, Congressional Research Service.

[7] International Energy Agency, *World Energy Outlook 2006*, p. 349.

[8] Energy Information Administration, at http://tonto.eia.doe.gov/dnav/pet/hist/wtotworldw.htm.

[9] EIA, http://www.eia.doe.gov/emeu/international/gasprice.html

[10] World Energy Outlook, *op. cit.*, pp. 139, 141.

[11] World Energy Outlook, *op. cit.*, p. 140.

[12] CRS Report RL33442, *Nuclear Power: Outlook for New U.S. Reactors*, by Larry Parker and Mark Holt.

[13] Interdisciplinary MIT Study, *The Future of Nuclear Power*, Massachusetts Institute of Technology, 2003, p. 79.

[14] "World Nuclear Performance in 2008 Close to Output in 2007," *Nucleonics Week*, March 5, 2009, p. 1.

[15] *Nuclear Engineering International*, November 2005, p. 37.

[16] Uranium Information Centre, *The Economics of Nuclear Power*, Briefing Paper 8, January 2006, p. 3.

[17] "U.S. Utility Operating Costs, 2008," *Nucleonics Week*, December 24, 2009.

[18] CRS Report RL34746, *Power Plants: Characteristics and Costs*, by Stan Mark Kaplan.

[19] Office of Technology Assessment, *Nuclear Power in an Age of Uncertainty*, OTA-E-216, February 1984, p. 59.

[20] Jan Willem Storm van Leeuwen and Philip Smith, *Nuclear Energy, the Energy Balance*, July 31, 2005, Chapter 3, p. 2.

[21] CRS Report RL34746, *Power Plants: Characteristics and Costs*, by Stan Mark Kaplan, Table B-3.

[22] World Nuclear Association, http://www.world-nuclear.org/info/reactors.html.

[23] International Energy Agency, *World Energy Outlook 2008*, pp. 508, 522.

[24] World Nuclear Association Reactor Database, at http://www.world-nuclear.org/reference/reactorsdb_index.php.

[25] World Nuclear Association, http://www.world-nuclear.org/info/reactors.html.

[26] IAEA, *Country Nuclear Fuel Cycle Profiles*, 2nd ed., 2005.

[27] U.S. DOE Energy Information Administration, http://www.eia.doe.gov/fuelnuclear.html.

[28] World Nuclear Association, http://www.world-nuclear.org/info/inf22.html.

[29] Note: Metric tons is the unit of measurement for uranium fuel. One metric ton is approximately 2,200 pounds. International Atomic Energy Agency, *Management of high enriched uranium for peaceful purposes: Status and trends* (IAEA-TECDOC-1452), June 2005.

[30] World Nuclear Fuel Market website, at http://www.wnfm.com/public/default.htm.

[31] Information on the Uranium Price Information System is available through NAC International at (678) 328-1211 or e-mail at gleamon@nacintl.com.

[32] Nine U.S. companies, 10 non-U.S. companies, 12 utilities, four producers, two traders, and one supplier.

[33] New York Mercantile Exchange, at http://www.nymex.com/UX_pre_agree.aspx.

[34] CME Globex is a global electronic trading platform for trading futures products. NYMEX ClearPort Clearing provides traders an interface where transactions are posted, margin requirements are calculated, and the transactions are processed by the clearinghouse.

[35] David Stellfox, "Industry Moves Closer to Having Standardized Uranium Contract," *NuclearFuel*, February 9, 2009.

[36] World Nuclear Association, *World Uranium Mining*, http://www.world-nuclear.org/info/inf23.html.

[37] World Nuclear Association, *Uranium Markets*, March 2007, at http://www.world-nuclear.org/info/inf22.html

[38] Ibid.

[39] Nuclear Energy Agency, *op. cit.*, p. 13.

[40] Ruthane Neely and Jeff Combs, "Diffusion Fades Away," *Nuclear Engineering International*, September 2006, p. 24.

[41] Thomas L. Neff, *The International Uranium Market*, Ballinger Publishing Co., 1984.

[42] EIA, Purchases of Enrichment Services by Owners and Operators of U.S. Civilian Nuclear Power Reactors by Origin Country and Year, http://www.eia.doe.gov/cneaf/nuclear/umar/table16.html.

[43] *USEC to Site American Centrifuge Plant in Piketon, Ohio—Technology Expected to Be World's Most Efficient for Enriching Nuclear Fuel*, at http://www.usec.com/v2001_02/Content/News/News Template.asp?page=/v2001_02/ Content/News/NewsFiles/01-12-04.htm.

[44] USEC Inc., "USEC Provides American Centrifuge Update," press release, November 2, 2009, http://www.usec.com/ NewsRoom/NewsReleases/USECInc/2009/2009-11-02-USEC-Provides-American-Centrifuge.htm.

[45] URENCO http://www.urenco.com/fullArticle.aspx?m=1371.

[46] URENCO http://www.urenco.com/fullArticle.aspx?m=1371.

[47] AREVA http://www.areva-nc.com/servlet/ContentServer?pagename= cogema_en%2FPage%2Fpage_site_ prod_ full_ template&c=Page&cid=1039482707079.

[48] NRC, *Uranium Enrichment*, http://www.nrc.gov/materials/fuel-cycle-fac/ur-enrichment.html. [49] http://www.usec.com.

[50] Ibid.

[51] IAEA, Country Nuclear Fuel Cycle Profiles, 2nd ed.

[52] Nuclear Engineering International, *2007 World Nuclear Industry Handbook*, p. 207.

[53] World Nuclear Association, *Processing of Used Nuclear Fuel for Recycle*, March 2007, at http://www.world-nuclear.org/info/inf69.html.

[54] World Nuclear Association, *Mixed Oxide Fuel (MOX)*, November 2006, at http://www.world-nuclear.org/info/inf29.html.

[55] World Nuclear Association, *Uranium Markets*, March 2007.

[56] Isotopes are atoms of the same chemical element but with different numbers of neutrons in their nuclei.

[57] The core of a breeder reactor is configured so that more fissile ^{239}Pu is produced from ^{238}U than the amount of fissile material initially loaded into the core that is consumed (^{235}U or ^{239}Pu). In a breeder, therefore, enough fissile material could be recovered through reprocessing to refuel the reactor and to provide fuel for additional breeders. The core of a fast reactor can also be configured to produce less ^{239}Pu than fissile material consumed, if the primary goal is to eliminate ^{239}Pu from spent fuel. In that case, much less ^{238}U ultimately would be converted to ^{239}Pu and therefore less total energy produced from a given amount of natural uranium.

[58] This section was prepared by Mary Beth Nikitin (Analyst in WMD Nonproliferation) in the Foreign Affairs, Defense, and Trade Division, and Anthony Andrews (Specialist in Energy Policy) in the Resources, Science, and Industry Division.

[59] "Towards a Safer World," at http://f40.iaea.org/worldatom/Press/Statements/2003/ebTE20031016.shtml.

[60] "Multilateral Approaches to the Nuclear Fuel Cycle: Expert Group Report submitted to the Director General of the International Atomic Energy Agency," February 22, 2005, (INFCIRC/640). Available at http://www.iaea.org/ Publications/ocuments/Infcircs/2005/infcirc640.pdf.

[61] Ibid., p. 98.

[62] International Atomic Energy Agency, *Possible New Framework for the Utilization of Nuclear Energy: Options for Assurance of Supply of Nuclear Fuel*, June 2007.

[63] Tariq Rauf, "Realizing Nuclear Fuel Assurances: Third Time's the Charm," Presentation to the Carnegie International Nonproliferation Conference, June 24, 2007, at http://www.carnegieendowment.org/files/fuel_assurances_rauf.pdf.

[64] "President Announces New Measures to Counter the Threat of WMD," February 11, 2004, at http://www.whitehouse.gov/news/releases/2004/02/20040211-4.html.

[65] "The Last Word: Sergei Kirienko," *Newsweek*, February 20, 2006 issue, at http://www.msnbc.msn.com/id/11299203/ site/newsweek/.

[66] "Concept for a Multilateral Mechanism for Reliable Access to Nuclear Fuel," Proposal as sent to the IAEA from France, Germany, the Netherlands, Russia, Ireland, and the United States, May 31, 2006, IAEA GOV/INF/2006/10.Available at http://www-pub.iaea.org/MTCD/Meetings/PDFplus/2006/cn147ConceptR A_ NF.pdf.

[67] WNA's report is available at http://www.world-nuclear.org/reference/pdf/security.pdf.

[68] *The Road to 2010: addressing the nuclear question in the 21st century*, July 2009, http://www.cabinetoffice. gov.uk/ reports/roadto2010.aspx

[69] INFCIRC/707, June 4, 2007.

[70] Nuclear Threat Initiative is a private organization founded in 2001 by Mr. Ted Turner and former Senator Sam Nunn. It is now classified as a 501(c)3 public charity.

[71] Nuclear Threat Initiative Commits $50 million to Create IAEA Nuclear Fuel Bank, International Atomic Energy Agency Press Release, September 19, 2006. Available at http://www.nti.org/c_press/ release_IAEA_ Fuelbank_091906.pdf.

[72] "Norway Contributes $5 Million to IAEA Nuclear Fuel Reserve," Norwegian Ministry of Foreign Affairs Press Release No. 027/08, February 27, 2008. http://www.regjeringen.no/en/dep/ud/Press-Contacts/News/2008/ fuelreserve.html?id=50210

[73] "UAE Commits $10 Million to Nuclear Fuel Reserve Proposal," IAEA Press Release, August 7, 2008 http://www.iaea.org/NewsCenter/News/2008/uaecontribution.html; "UAE Commitment Gives NTI/IAEA Fuel Bank Critical Momentum," NTI Press Release, August 7, 2008. http://www.nti.org/c_press/ release_UAE%20fuel%20bank%2080708.pdf

[74] "Sam Nunn Applauds EU Contribution to IAEA Fuel Bank," NTI Press Release, December 10, 2008. http://www.nti.org/c_press/statement_Nunn_EU_fuel_bank_121008.pdf

[75] "Multinational Fuel Bank Reaches Key Milestone," IAEA Staff Report, March 6, 2009. http://www.iaea.org/ NewsCenter/News/2009/fbankmilestone.html

[76] For more detailed treatment of these questions see, New Framework for the Utilization of Nuclear Energy in the 21st Century: Assurances of Supply and Nonproliferation, IAEA Special Event, Speech by Laura Holgate, September 19, 2006, at http://www.nti.org.

[77] Sylvia Westall, "Obama-backed nuclear fuel bank plan stalled at IAEA," *Reuters*, June 18, 2009.

[78] H.Rept. 110-477 to accompany H.R. 1585.

[79] "U.S. Donates $50 million for the IAEA International Nuclear Fuel Bank," *NNSA Press Release*, August 4, 2008. http://nnsa.energy.gov/news/2090.htm

[80] Division C, Title III, Section 312 of the FY2008 Consolidated Appropriations Act rescinded 1.6 percent of discretionary budget authority for Congressionally directed projects, this includes the fuel bank. http://www.whitehouse.gov/omb/legislative/fy08consolidated_reductions_01_25_08.pdf

[81] "Board of Governors Approves Plan for Nuclear Fuel Bank: Russian Plan to Supply Low-Enriched Uranium," IAEA Staff Report, November 27, 2009, http://www.iaea.org/NewsCenter/News/2009/nuclfuelbank.html

[82] "Development of the initiative of the Russian Federation to establish a reserve of low enriched uranium for the supply of LEU to the International Atomic Energy Agency for its member states," Working Paper of the Russian Federation, Preparatory Committee for the 2010 NPT Review Conference, May 6, 2009, NPT/CONF.2010/PC.III/WP.25

[83] "DOE/NNSA Reliable Fuel Supply Gains Momentum," NNSA Press Release, November 7, 2006.

[84] "News Analysis: The Growing Nuclear Fuel-Cycle Debate," Arms Control Today, November 2006. Available at http://www.armscontrol.org/act/2006_11/NAFuel.asp.

[85] See also CRS Report RS22937, *Nuclear Cooperation with Other Countries: A Primer*, by Paul K. Kerr and Mary Beth Nikitin.

[86] "NNSA Awards Contract for Reliable Fuel Supply Program," NNSA Press Release, June 29, 2007.

[87] "NNSA Announces Contract to Downblend 12 Metric Tons of Surplus Highly Enriched Uranium," NNSA Press Release, June 23, 2009.

[88] Anya Loukianova, "The International Uranium Enrichment Center at Angarsk: A Step Towards Assured Fuel Supply," NTI Issue Brief, updated November 2008, http://www.nti.org/e_research/e3_93.html

[89] A Russian-Kazakh joint venture Uranium Enrichment Center (distinct from the IUEC) is also located at Angarsk.

[90] Rosatom website, "Uranium Enrichment Division," accessed February 22, 2010. http://www.rosatom.ru/en/ energy_complex/uranium_enrichment/

[91] "Russia's Angarsk international enrichment center open for business," *Nuclear Fuel*, September 24, 2007.

[92] "Communication received from the Resident Representative of the Russian Federation to the IAEA on the Establishment, Structure and Operation of the International Uranium Enrichment Centre," INFCIRC/708, June 8, 2007.

[93] Russia, as a nuclear weapon state under the NPT, has a voluntary safeguards agreement that allows, but does not require, inspections.

[94] "Safe Enrichment for All," Handelsblatt newspaper, May 2, 2007, in English at https://www.diplo.de/diplo/en/ Infoservice/Presse/Interview/2007/070502-Handelsblatt.html.

[95] DOE Global Nuclear Energy Partnership home page, at http://www.gnep.energy.gov.

[96] Richard Garwin, "R&D Priorities for GNEP," Testimony to House Science Committee, April 6, 2006.

[97] Frank von Hippel, "GNEP and the U.S. Spent Fuel Problem," Briefing for Congressional Staff, March 10, 2006, at http://www.princeton.edu/~globsec/publications/pdf/HouseBriefing10March06rev2.pdf; Frank von Hippel, "Managing Spent Fuel in the United States: The Illogic of Reprocessing," International Panel on Fissile Materials, January 2007, at http://www.fissilematerials.org/ipfm/site_down/ipfmresearchreport03.pdf.

[98] "DOE's Spent Nuclear Fuel Reprocessing R&D Program Should Be Scaled Back; Boosted Efforts to Get New Nuclear Power Plants Online Needed," National Academies News Release, October 27, 2007, http://www8.nationalacademies.org/onpinews/newsitem.aspx?RecordID=11998.

[99] "Remarks as Prepared for Delivery by U.S. Secretary of Energy Samuel W. Bodman," 2nd Global Nuclear Energy Partnership Ministerial Opening Session Vienna, Austria, September 16, 2007.

[100] Global Nuclear Energy Partnership, http://www.gneppartnership.org/index.htm.

[101] See CRS Report RL33016, *U.S. Nuclear Cooperation with India: Issues for Congress*, by Paul K. Kerr.

[102] "Black box" or turn-key plants would be built so that recipients would not be able to replicate the facilities, including sensitive components.

[103] Daniel Horner, Mark Hibbs, "G8 adopts interim measure on sensitive nuclear exports," *Nucleonics Week*, July 17, 2008.

[104] See paragraph 66 of the Hokkaido Toyako G-8 Summit Leaders Declaration, July 8, 2008, http:// www.mofa.go.jp/ policy/economy/summit/2008/doc/doc080714__en.html.

[105] See paragraph 8 of the L'Aquila G-8 Summit's Statement on Nonproliferation, July 2009, http:// www.g8italia2009.it/static/G8_Allegato/2._LAquila_Statent_on_Non_proliferation.pdf

[106] The IAEA proposal is "Multilateral Approaches to the Nuclear Fuel Cycle: Expert Group Report Submitted to the Director General of the International Atomic Energy Agency," INFCIRC/640, International Atomic Energy Agency, February 22, 2006, p. 18. Available at http://www.iaea.org/Publications/ Documents/Infcircs/ 2005/infcirc640.pdf.The Six-Country Concept is "Concept for a Multilateral Mechanism for Reliable Access to Nuclear Fuel," Proposal as sent to the IAEA from France, Germany, the Netherlands, Russia, Ireland, and the United States, May 31, 2006. Available at http://www-pub.iaea.org/MTCD/Meetings/ PDFplus/2006/cn147_ConceptRA_NF.pdf.

[107] For an analysis of these past proposals, see Lawrence Scheinman, "Equal Opportunity: Historical Challenges and Future Prospects of the Nuclear Fuel Cycle," *Arms Control Today*, May 2007.

[108] See CRS Report RL33558, *Nuclear Energy Policy*, by Mark Holt.

[109] See CRS Report RS22937, *Nuclear Cooperation with Other Countries: A Primer*, by Paul K. Kerr and Mary Beth Nikitin.

[110] CRS Report RL33016, *U.S. Nuclear Cooperation with India: Issues for Congress*, by Paul K. Kerr

[111] CRS Report R40344, *The United Arab Emirates Nuclear Program and Proposed U.S. Nuclear Cooperation*, by Christopher M. Blanchard and Paul K. Kerr.

In: International Policies and Issues of Nuclear Safety… ISBN: 978-1-61761-596-2
Editor: Matthew D. Leeland

Chapter 2

NUCLEAR SAFETY: CONVENTION ON NUCLEAR SAFETY IS VIEWED BY MOST MEMBER COUNTRIES AS STRENGTHENING SAFETY WORLDWIDE

United States Government Accountability Office

WHY GAO DID THIS STUDY

Currently, 437 civilian nuclear power reactors are operating in 29 countries, and 56 more are under construction. After the Chernobyl accident, representatives of over 50 nations, including the United States, participated in the development of the Convention on Nuclear Safety, a treaty that seeks to promote the safety of civilian nuclear power reactors. The Convention has been in force since 1996. GAO was asked to assess (1) parties' views on the benefits and limitations of the Convention, (2) efforts to improve implementation of the Convention, and (3) how International Atomic Energy Agency (IAEA) programs complement the Convention's safety goals. GAO surveyed the 64 parties to the Convention for which it was in force at the time of GAO's review and analyzed the responses of the 32 that completed it, analyzed relevant documents, and interviewed U.S. and foreign officials.

WHAT GAO RECOMMENDS

GAO recommends, among other things, that the Department of State, in coordination with NRC, work with other parties to the Convention to encourage the use of performance metrics in national reports to track progress toward improving safety of civilian nuclear power plants and expand efforts to increase the number of reports posted to IAEA's public Web site. The Department of State generally agreed with these recommendations. NRC generally agreed with GAO's report but did not specifically agree or disagree with these recommendations.

What GAO Found

The Convention on Nuclear Safety plays a useful role in strengthening the safety of civilian nuclear power reactors worldwide, according to most parties to the Convention that responded to GAO's survey and representatives of parties GAO interviewed. In particular, parties indicated that the Convention's obligations to (1) establish effective legislative and regulatory frameworks and strong, independent nuclear regulatory bodies and (2) prepare a national report every 3 years that describes the measures the country has taken to achieve the Convention's nuclear safety goals, are among its most useful contributions. The countries present their national reports at review meetings, address questions that may arise about the reports, and assess and ask questions about the reports of other parties. This is known as the peer review process. Some concerns were raised about limited public access to Convention proceedings, some countries' lack of resources to fully participate in the review meetings, and the absence of performance metrics in the national reports to gauge progress toward meeting safety goals and objectives. Half of the parties responding to GAO's survey stated that the lack of performance metrics limited the usefulness of the Convention. Neither the Department of State nor the Nuclear Regulatory Commission (NRC) has formally proposed the adoption of performance metrics. However, NRC officials told GAO that performance metrics could be useful. In addition, the number of parties posting their national reports to IAEA's public Web site has declined since 2005. NRC and Department of State officials told GAO that the United States has always made its national report available on the Internet. However, the U.S. approach has been to lead by example rather than taking an active role in encouraging other parties to post their reports. Further, universal participation would advance achievement of the Convention's goals. Several representatives from countries who are parties to the Convention told GAO that Iran should ratify the Convention. In their view, without Iran's participation, the international community has limited or no insight on, or access to, Iran's civilian nuclear power program. Russia, which is helping Iran build the nuclear reactor at Bushehr, may condition continued assistance on Iran becoming a party to the Convention, according to Russian officials.

The parties have taken some actions to improve the Convention's implementation, and more proposals are being considered. Steps have been taken to make the process for asking questions during peer review meetings more open and to increase the amount of time available for preparing for the review meetings.

IAEA nuclear safety programs, which predate the Convention, complement the Convention's safety goals through the Technical Cooperation program, safety standards, and peer review missions. The Technical Cooperation program supports, among other things, the development of nuclear power. IAEA has established nuclear safety standards and also promotes nuclear safety through peer review missions that evaluate the operations of a member state's nuclear regulatory system and nuclear power plant operational safety.

Abbreviations

DOE	Department of Energy
ENSREG	European Nuclear Safety Regulators Group

EU	European Union
IAEA	International Atomic Energy Agency
INPO	Institute of Nuclear Power Operations
IRRS	Integrated Regulatory Review Service
FSA	Freedom Support Act
NEA	Nuclear Energy Agency
NRC	Nuclear Regulatory Commission
NSCI	Nuclear Safety Cooperation Instrument
OSART	Operational Safety Review Team
SEED	Support for Eastern European Democracies
TACIS	Technical Assistance to the Commonwealth of Independent States
TC	Technical Cooperation
WANO	World Association of Nuclear Operators
WENRA	Western European Nuclear Regulators' Association

April 29, 2010
The Honorable Daniel K. Akaka
Chairman
Subcommittee on Oversight of Government
Management, the Federal Workforce, and the District of Columbia
Committee on Homeland Security and Governmental Affairs
United States Senate

Dear Mr. Chairman:

There are currently 437 civilian nuclear power reactors operating in 29 countries, generating about 14 percent of the world's electricity, and 56 more nuclear power reactors are currently under construction. The safe operation of nuclear power reactors worldwide has been a long-standing concern of the international community. In the aftermath of the Chernobyl accident, representatives of over 50 nations, including the United States, participated in the development of the Convention on Nuclear Safety (the Convention), a multilateral treaty that seeks to strengthen the safety of civilian nuclear power reactors.[1] Established in the mid-1990s, the Convention seeks to achieve its safety objectives through countries' adherence to general safety principles rather than through technical standards. Officials describe the Convention as incentive-oriented, designed to maximize the number of countries that will support and sign it, with the goal of making it acceptable and useful to countries with potentially unsafe power reactors in Eastern Europe and the countries of the former Soviet Union. According to U.S. officials, the main purpose of the Convention is to get these countries, as well as developing nations, to make commitments to improve their reactors and develop a safety culture.[2]

Currently, 65 countries and 1 international organization are parties to the Convention, including all countries that currently operate civilian nuclear power reactors.[3] For the purpose of this chapter, we refer to countries that have ratified, accepted, or approved the Convention as parties. The United States ratified the Convention in 1999.

The Convention calls on parties to, among other things, (1) establish and maintain a legislative framework and an independent regulatory body to govern the safety of nuclear installations; (2) establish procedures to ensure that technical aspects of safety, such as the siting, design, construction, and operation of nuclear power reactors, are adequately considered; (3) maintain an acceptable level of safety throughout the life of the installations by, for example, considering safety to be a priority and establishing a quality assurance program; and (4) prepare and routinely test emergency plans. The Convention does not impose sanctions when countries do not follow these safety principles.

Under the terms of the Convention, each country—regardless of whether it operates nuclear power plants or not—is required to submit a national report that identifies the measures taken to implement each of the nuclear safety obligations contained in the Convention. Obligations cover such points as siting, design, construction, and operation of civilian nuclear power installations. The parties to the Convention have also established detailed guidance to help parties prepare their national reports. The purpose of the guidance is to encourage parties to describe the steps they are taking to meet the Convention's obligations and to facilitate other parties' review of the national reports of other countries. The countries meet every 3 years in Vienna, Austria, to present their national report, address questions that may arise about the report, and assess and ask questions about the reports of other parties.[4] This is known as the peer review process, and it is considered central to the Convention's success because it is the means by which the parties assess the steps being taken to meet safety obligations. As part of this peer review process, countries meet in six groups composed primarily on the basis of the number of reactors that each country operates. This process ensures that the six countries with the most reactors—the United States, France, Japan, Russia, South Korea, and the United Kingdom—are never in the same group. Within this confidential group setting, all member countries have the opportunity to examine and review what each country reports it is doing to meet its nuclear safety obligations. These meetings are hosted by the International Atomic Energy Agency (IAEA), which serves as the Convention's secretariat and provides administrative support.[5] To date, four review meetings have taken place, and the fifth is scheduled for April 2011.

The Convention has taken on increased significance in recent years as countries are either expanding their existing nuclear power capacity or planning to establish new programs. In 2009, IAEA estimated that by 2030 the world's capacity for nuclear electricity production will significantly increase. Most of this increase in capacity is expected to occur in countries that have established civilian nuclear power programs, such as China, Japan, and South Korea. China, for example, has announced its intention to spend $50 billion to build 32 new nuclear plants by 2020 and currently has 21 plants under construction. Both India and Pakistan are also moving forward with plans to significantly increase their production of nuclear power, building plants that will more than double their production of nuclear energy in the next decade.

In addition, countries such as Jordan and the United Arab Emirates, which do not yet have civilian nuclear power programs, are actively moving to build the necessary regulatory infrastructure for such programs as they explore agreements with the world's leading nuclear reactor vendors. The United Arab Emirates, for example, recently signed a $20 billion agreement with a consortium of South Korean vendors to begin construction of four 1,400-megawatt nuclear power reactors in 2012. Other countries, such as Indonesia, Libya, Thailand, and Vietnam, have expressed their intent to build civilian nuclear power plants. Still others, such as Algeria, Belarus, Egypt, Nigeria, and Yemen, are considering moving forward with civilian nuclear power programs.

To assist Congress in its deliberations in the past, we identified some limitations of the Convention.[6] Specifically, we noted that (1) public access to the peer review process was unclear and (2) the effectiveness of the peer review process was uncertain because of concerns about how well the country groups formed for peer review meetings would function. We also pointed out that the Convention lacked an enforcement mechanism.

Now that the Convention has been in force for more than a decade, you asked us to evaluate the extent to which it is achieving its primary goal: promoting the safe operation of civilian nuclear power reactors worldwide. Accordingly, we assessed (1) parties' views on the perceived benefits and limitations of the Convention, (2) efforts to improve the implementation of the Convention, and (3) how IAEA programs complement the Convention's safety goals and objectives.

To assess parties' views on the perceived benefits and limitations of the Convention for improving the safety of civilian nuclear power reactors worldwide, we administered a Web-based survey—which can be viewed at GAO-10-550SP—to 64 parties to the Convention and analyzed the responses of the 32 that completed it.[7] This chapter does not contain all the results from the survey. To assess the potential for nonresponse bias in our survey results, we compared selected characteristics of nonresponding countries, such as (1) length of time as a party to the Convention, (2) nuclear power status and number of nuclear power plants, (3) region, (4) countries that operate Soviet-designed reactors, and (5) European Union (EU) membership, to those of the responding parties. The distribution of these characteristics among responding and nonresponding parties was well-balanced. To encourage respondents to complete the survey, we sent an e-mail reminder to each nonrespondent about 2 weeks after our initial e-mail message and followed up with additional e-mails and telephone calls. Additionally, to encourage honest and open responses, in the introduction to the survey, we pledged that we would report information in the aggregate and not report data that could identify a particular respondent. We also interviewed representatives of 17 nuclear and nonnuclear parties to the Convention, including officials from the Nuclear Regulatory Commission (NRC) and the Department of State (State) who represent the United States at the Convention. Of the 17 we interviewed, 9 completed the survey, and 8 did not. In total, we obtained the views of 40 parties to the Convention. We also analyzed various Convention-related documents from NRC and State as well as from IAEA and the EU. To assess efforts to improve the implementation of the Convention, we reviewed Convention documents and interviewed NRC and State officials who have attended Convention organizational, working group, and review meetings where such efforts have been discussed. To assess the extent to which IAEA programs complement the Convention's safety goals and objectives, we analyzed, among other things, Convention minutes of meetings and rules of procedure. We also interviewed IAEA officials, U.S. officials at the U.S. Missions in Vienna and Brussels, and the representatives of 17 parties to the Convention. To determine the cost to the United States to participate in the Convention and IAEA's costs to support the Convention for one 3-year cycle, we analyzed budget information from NRC, State, and IAEA. We also assessed the reliability of the data we obtained and interviewed knowledgeable NRC, Department of Energy (DOE), State, EU, and IAEA officials on the reliability of the data. We determined that these data were sufficiently reliable for the purposes of this chapter. Appendix III explains our methodology in greater detail.

We conducted this performance audit from February 2009 to April 2010, in accordance with generally accepted government auditing standards. Those standards require that we plan and perform the audit to obtain sufficient, appropriate evidence to provide a reasonable basis for our findings and conclusions based on our audit objectives. We believe that the evidence obtained provides a reasonable basis for our findings and conclusions based on our audit objectives.

BACKGROUND

The Convention is one of a number of cooperative efforts by the international community to improve nuclear safety worldwide and is meant to complement these other efforts. For example, as we previously reported, the United States and 20 other countries and international organizations contributed $1.9 billion to improve nuclear safety in countries operating Soviet-designed nuclear reactors.[8] The United States alone has spent over $770 million since the Chernobyl accident on nuclear safety assistance to Russia, Ukraine, Kazakhstan, Armenia, and several other countries through DOE and NRC programs. According to an agency official, DOE's nuclear safety assistance programs have focused on physical safety enhancements to Soviet-designed reactors, while NRC has worked to increase the capacity and stature of recipient countries' regulatory bodies to ensure the continuing operational safety of such reactors. In addition, a separate fund was established to help stabilize the damaged reactor at Chernobyl by constructing a new containment structure. As we reported, the estimated cost of this effort was $1.2 billion as of 2007, of which the United States pledged $203 million.[9] Since 1991 the EU has spent over $1.9 billion on international nuclear safety assistance. See appendix II for more information about U.S. and EU expenditures to promote international nuclear safety. These expenditures are not used to support the implementation of the Convention. Matters pertaining to U.S. financial support to the Convention are contained on page 28 of this chapter.

In addition to the Convention, other multilateral organizations—the Nuclear Energy Agency (NEA),[10] the Western European Nuclear Regulators' Association (WENRA),[11] the European Nuclear Safety Regulators Group (ENSREG),[12] and the EU—are making efforts to advance the safety of civilian nuclear power. All member or observer countries of the NEA, WENRA, ENSREG, and the EU are also parties to the Convention. The NEA, for example, has created several specialized committees to facilitate exchanges of technical information and to organize joint research projects to improve national safety practices. WENRA works to develop common approaches to nuclear safety among the chief nuclear regulators in Europe. ENSREG, among other things, aims to maintain and continuously improve the safety of nuclear installations in the EU. In June 2009, the EU adopted a directive creating a framework for (1) maintaining and promoting the continuous improvement of nuclear safety and its regulation and (2) ensuring that EU member states provide a high level of nuclear safety to protect workers and the public against radiation from nuclear installations. This framework is based in part on IAEA safety documents and the obligations of the Convention. EU members are required to incorporate the directive into their national legislation by June 2011.

Other conventions have been established to advance international nuclear safety and are administered by IAEA's Department of Safety and Security. Two "emergency conventions" obligate parties to provide early notification of a nuclear accident and to render assistance in the event of such an accident or a radiological emergency, and two other conventions obligate parties to safely manage spent fuel and radioactive waste and to take effective action to physically protect nuclear material.

THE MAJORITY OF PARTIES WE SURVEYED AND INTERVIEWED REPORTED THAT THE CONVENTION HAS STRENGTHENED NUCLEAR SAFETY WORLDWIDE

The Convention on Nuclear Safety has played a useful role in strengthening the safety of civilian nuclear power reactors worldwide, according to most survey respondents and representatives of parties to the Convention we interviewed. In their view, efforts to improve parties' nuclear regulatory capabilities and the obligation to prepare a national report every 3 years are among the most useful contributions the Convention has made to increased nuclear safety. In addition, parties responded that the Convention has promoted opportunities for communication and promoted sharing of useful technical information about nuclear safety. According to most parties we surveyed and interviewed, maintaining confidentiality about the safety issues discussed was key to the success of the peer review process. Despite the Convention's positive impacts on nuclear safety, some parties have concerns about limited public access to the Convention's proceedings, some parties' limited resources to fully participate in Convention activities, and the absence of metrics to assess progress toward meeting safety goals.

The Convention Has Strengthened Nuclear Safety by Promoting Improved Regulatory Capabilities and Requiring National Reports

Nearly all parties responding to our survey reported that the Convention has been very useful or somewhat useful in helping to strengthen nuclear safety both in their country and worldwide. In all, these parties operate 404—or more than 92 percent—of the world's 437 operating civilian nuclear power reactors. In addition, we also interviewed representatives from IAEA member states, nuclear regulatory organizations, and the EU (17 in all) who expressed similar views about the Convention. Survey respondents and parties we interviewed identified several Convention obligations as having helped strengthen the safety of civilian nuclear power programs. The obligations cited most frequently were (1) establishing an effective legislative and regulatory framework (laws and regulations) and a strong, effective, and independent nuclear regulatory body[13] and (2) preparing a national report every 3 years that describes the measures the country has taken to achieve the Convention's safety goals.

In addition, some of the 17 parties we interviewed stated that the Convention has contributed to and promoted the independence and effectiveness of their country's nuclear regulatory bodies. For example, an Austrian nuclear regulator told us he thought that this promotion of effective regulatory capacity is one of the Convention's greatest contributions to international nuclear safety. Moreover, representatives of China and Pakistan told us that the Convention was influential in leading their countries to increase the independence and effectiveness of their nuclear regulators. NRC officials expressed a similar view, noting that parties to the Convention have taken many steps to develop more effective laws and regulations and increase the capacities and independence of their nuclear regulators.

The requirement to prepare a national report describing the steps parties have taken to meet the Convention's nuclear safety obligations also plays a large role in strengthening the safety of civilian nuclear power programs, according to survey respondents. Almost all survey respondents indicated that the presentation of national reports in country groups was a very or somewhat effective way for sharing best safety practices. Most survey respondents reported that preparing the national report has either greatly or somewhat improved opportunities to examine their country's civilian nuclear power program. A number of parties we interviewed also said that this national report has been helpful in strengthening nuclear safety worldwide. NRC officials told us one effect of a national report is that nuclear regulators and plant operators are forced to think about even routine safety procedures and policies because the reports will be scrutinized by their peers. For example, as a result of questions raised by other parties on the national report prepared for the 2008 review meeting, the United States agreed to discuss with state governments and NRC licensees the benefits and costs of adopting stricter standards for protecting nuclear power plant workers and the public from exposure to radiation.

In our survey, we also asked some additional questions about parties' perceptions about how the peer review process affected the preparation of the 2008 reports. Specifically, among other things, we asked how likely parties thought reports were to include (1) comprehensive, detailed descriptions of measures taken to strengthen safety; (2) evidence that safety issues discussed in one review meeting were revisited in the next meeting and that the actions taken to address the issues were discussed in sufficient detail for parties to evaluate whether the safety concerns had been adequately addressed; and (3) sufficient technical detail to understand specific safety concerns. In each case, most survey respondents indicated that they thought reports were very or somewhat likely to include such information. We also asked how effectively the peer review process encouraged parties to provide detailed information in their 2008 national reports. Overall, most survey respondents indicated that the peer review process was very or somewhat likely to encourage parties to include detailed, comprehensive, and accurate information in their national reports.

According to Parties We Surveyed and Interviewed, the Convention Has also Improved Communication and Promoted Sharing of Technical Information about Nuclear Safety Issues

According to both survey respondents and parties we interviewed, the Convention has increased communication and encouraged the sharing of technical information to improve nuclear safety worldwide. There was wide agreement among the survey respondents that the Convention has improved communication among nuclear regulators; nuclear power plant operators; and other national organizations involved in the civilian nuclear power industry, such as, in the case of the United States, the Institute of Nuclear Power Operations (INPO).[14] More than half of the respondents to our survey indicated the Convention had "greatly" improved communication about safety issues affecting civilian nuclear power reactors. Most respondents to our survey agreed that the Convention had improved opportunities for sharing technical solutions to improve safety, such as reactor design improvements or fire safety enhancements. Russian and Ukrainian officials we spoke to provided examples of how the Convention has led to the sharing of nuclear safety information. Following are some examples:

- Russian nuclear regulatory officials told us that the Convention has played a useful role in promoting technical solutions to problems shared by countries operating similar types of reactors. Specifically, Russia and Finland have been developing a system to improve communication between their plant operators based on discussions that began with contacts made at Convention review meetings.
- A Ukrainian official told us his country's participation in the Convention has increased other countries' awareness of the safety problems confronting Ukraine's aging Soviet-designed nuclear reactors. He further noted that the Convention is one of many forums that Ukraine participates in that supports the strengthening of nuclear safety.

Confidentiality among the Parties to the Convention Has Been Key to the Success of the Peer Review Process

According to most parties we surveyed and interviewed, maintaining the confidentiality of information obtained during the Convention's meetings is critical to the peer review process. Most party representatives we spoke with agree that confidentiality should be preserved. For example, when asked if the public should be allowed to directly observe review meetings—and thereby gain direct access to a party's national report and any concerns or questions raised about it by other parties—approximately two-thirds of survey respondents said the public probably or definitely should not be given such access. Some parties we interviewed told us that, as a result of the confidentiality of the peer review process, their country's national reports have become more comprehensive. Three-quarters of survey respondents indicated that the quality of national reports prepared for review meetings had improved in the past 10 years.

Concerns Exist about Some Aspects of the Convention's Implementation

While the parties' perceptions of the value of the Convention are generally very positive, some concerns were raised about the lack of information provided to the general public about the Convention's proceedings, some countries' lack of resources to fully participate in the review meetings, and the absence of performance metrics. In addition, parties emphasize that without the participation of all countries with nuclear power programs in the Convention, the international community will have limited access and insight into countries'—such as Iran—civilian nuclear power programs.

Limited Public Access to the Convention's Proceedings

Notwithstanding the general agreement that preserving the confidentiality of the peer review process is important, most parties responding to our survey would like to see more public access to the results of review meetings. We have testified that, according to some experts familiar with international agreements that rely primarily on peer review, the public dissemination of information about parties' progress in meeting the terms of the Convention can play a key role in influencing compliance with the Convention's nuclear safety obligations.[15] Currently, only summary information of the peer review meeting is released to the public. This summary provides a brief introduction containing background on the Convention, an overview of the review process, and a synopsis of what the parties agree were the most important points discussed at the meeting. For example, the public report on the fourth review meeting, which took place in 2008, briefly summarizes the discussions of the parties on many topics discussed at the meeting, including parties' efforts to meet the challenges of maintaining adequate staffing and competence levels and ongoing concerns about the degree of independence of some parties' regulatory bodies. Any further details about any party's national report or questions and answers on the report remain confidential unless the party voluntarily releases it.

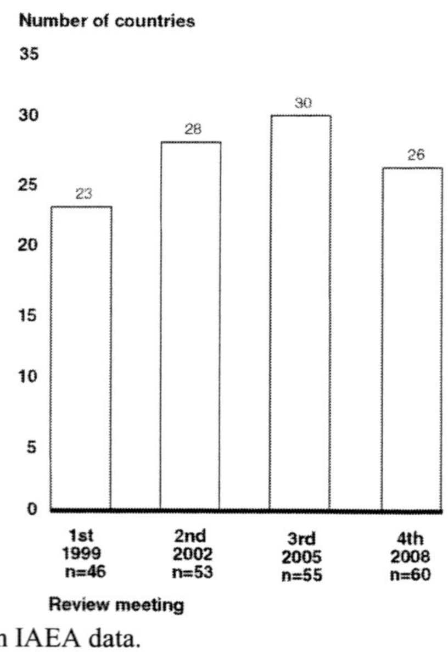

Source: GAO analysis based on IAEA data.

Note: This figure, with n = the number of parties that were obligated to submit a national report for that review meeting, includes national reports from all parties to the Convention that submitted them, regardless of their nuclear power status. All parties to the Convention are required to submit a national report for peer review. Parties that do not operate nuclear power plants, such as Austria, submit reports focusing, among other things, on the steps they have taken to prepare and test emergency plans to deal with an accident in a neighboring country that operate a nuclear power plant. Other nonnuclear countries may be considering establishing nuclear power programs and it is important for them to provide information in their national reports about the steps they are taking to meet the Convention's obligations including, for example, reactor design and siting requirements.

Figure 1. Number of Countries Posting National Reports on IAEA's Public Web Site, 1999-2008

French officials in particular have expressed an especially strong view regarding public access to information about the Convention's proceedings. In July 2009, in written responses to our questions, French officials stated that parties to the Convention should consider making the opening and closing sessions of review meetings open to the media. Further, a Norwegian official we spoke with suggested that some nongovernmental organizations should be allowed to attend review meetings as observers.

One way that some parties have attempted to increase public access to the Convention's proceedings is by posting their national reports and answers to written questions received on their national reports to IAEA's public Web site. While the number of parties to the Convention making their national reports available in this way has increased since the first review meeting was held in 1999, it has not increased significantly in several years and actually declined between the third review meeting in 2005 and the fourth review meeting in 2008. Specifically, 26 parties—about 43 percent of the 60 parties for whom the Convention had come into force by the due date for submitting the national report—posted their national report prepared for the 2008 review meeting. This was down from the 30 parties—or about 55 percent of parties to the Convention— posting reports prepared for the 2005 review meeting. In fact, eight countries that posted their national reports prepared for the 2005 review meeting—Argentina, Belgium, Bulgaria, Ireland, Japan, Latvia, the Slovak Republic, and South Korea—did not do so for the report prepared for the 2008 review meeting. However, three parties posted their national reports for the first time in 2008—Estonia and India, which had recently become parties to the Convention, and Pakistan, which became a party in the 1990s. Figure 1 shows the number of countries that posted their national reports to the IAEA public Web site for the four review meetings held thus far.

Officials from NRC and State told us that the United States has always made its national report available on the Internet. However, the U.S. approach has been to lead by example rather than taking an active role in encouraging other parties to the Convention to post their national reports to the Internet. IAEA officials told us it was important for parties to make as much information about their civilian nuclear power programs accessible as possible, but that it was for each party to determine how much information should be made public and how much should remain confidential. In addition to its public Web site, IAEA also maintains a secure, members-only Web site where parties are encouraged to post their national reports. According to NRC officials, parties have improved their participation in posting their reports to this Web site. Parties posted 17, 22, 57, and 61 national reports in 1999, 2002, 2005, and 2008, respectively.

The overwhelming majority of parties have never posted their answers to written questions about their nuclear power programs to the IAEA public Web site. The written questions and answers provide a great deal of information about each country's nuclear power program. According to an IAEA official, over 4,000 questions were prepared for the 2008 review meeting, and almost all were answered. As figure 2 shows, 3 countries posted these questions and answers to the IAEA public Web site for the first review meeting in 1999. While 11 countries posted questions and their answers to the IAEA's public Web site for the second review meeting, including the United States, 6 did so for the third review meeting, and 5 did so for the 2008 meeting. Only Slovenia and Switzerland—both nuclear power countries—have posted these questions and answers for all four meetings; the United Kingdom and Canada— the sixth and eighth largest nuclear power countries as measured by the number of operating reactors, respectively—have done so since 2002. The United States had not posted its answers to written questions received on its national report to IAEA's public Web site since 2002, although NRC officials stated that they have always posted them to the NRC Web site. We also found that other nuclear power countries such as Finland, Germany, Japan, and Spain have not posted their answers to written questions to the IAEA's public Web site since 2002, either. In 2008, Luxembourg became the first, and thus far only, nonnuclear party to post the answers to questions it received on its national report. Luxembourg's responses focused primarily on how it would respond to a nuclear accident in a neighboring country.

We met with NRC officials on March 15, 2010, to discuss an early draft of this chapter. At that time, we informed them that their answers to written questions on U.S. national reports were not available on IAEA's public Web site. NRC officials acknowledged that these responses were not readily accessible and said they would take steps to post them. On March 17, 2010, NRC informed us of the availability of their responses, and we verified that they were now on IAEA's public Web site.

Lack of Resources to Fully Participate in the Convention's Review Meetings

Some respondents to our survey reported the lack of resources to fully participate in the review meetings. Specifically, almost half of the survey respondents—ranging from parties with well-established civilian nuclear power programs to those with no nuclear power programs—report that a lack of resources has limited their country's ability to develop their national report. As we noted in our March 1999 testimony,[16] NRC officials anticipated this lack of staff resources and/or travel money could be a problem. We reported that NRC officials told us that, because of differences in parties' nuclear safety programs and available resources, they anticipated unevenness in the quality and detail of some national reports. In addition, half of the parties responding to our survey reported that a lack of resources has limited their ability to attend review meetings, and more than three-quarters indicated that a lack of resources has inhibited their ability to send representatives to all of the country group meetings. According to NRC officials, this is important because the country groups meet simultaneously, and it is in these meetings where the national reports are presented and questions about them are addressed.

Not being able to attend country group meetings reduces opportunities to learn from other parties' nuclear safety experiences. In addition, NRC officials recently told us that since much of the peer review of national reports can occur in the 7 months before the review meeting, limited resources may reduce the ability of some parties to take full advantage of this opportunity. That is, according to NRC officials, some countries do not have the staff resources to devote to preparing for review meetings by reading national reports, formulating and submitting written questions, and reviewing the parties' written responses to the written questions.

Lack of Performance Metrics to Gauge Progress in Strengthening Safety

The Convention does not include performance metrics to gauge its impact on improving safety. As a result, it provides no systematic way to measure where and how progress in improving safety in each country has been made. During the course of this review, we asked parties if the lack of performance metrics limited the usefulness of the Convention. Half the parties responding to our survey indicated that it did. Performance indicators and benchmarks are currently being used to track safety in civilian nuclear power plants that could be adapted to help countries enhance safety. For example, the World Association of Nuclear Operators (WANO)[17] publishes quantitative indicators of nuclear plant performance for 11 key areas, including industrial safety accidents and unplanned automatic shutdowns of nuclear power plants. Although the Convention itself lacks performance metrics, one-quarter of parties responding to our survey reported that they themselves measure progress toward Convention goals using performance metrics—specifically, in some cases, by comparing their activities with the results of IAEA safety review missions to countries that request them and actions taken in response to questions and comments from other parties at Convention review meetings.

Neither State nor NRC has formally proposed the adoption of performance metrics. However, NRC officials told us that performance metrics could play a useful role in helping parties measure their progress toward meeting safety obligations and that they could be introduced through a modification to the rules and procedures governing the Convention. Specifically, Article 22 of the Convention provides for the preparation of guidelines by the parties regarding the form and structure of their national reports. The guidelines can be revised by consensus at review meetings. The guidelines provide only suggestions for drafting the reports; parties remain free to structure their reports as they see fit. However, the suggestions provided are very detailed and touch upon more than just form and structure. For example, the guidelines provide detailed suggestions on the content of the national reports. They also contain an appendix detailing voluntary practices that parties are encouraged to engage in regarding the public availability of their national reports.

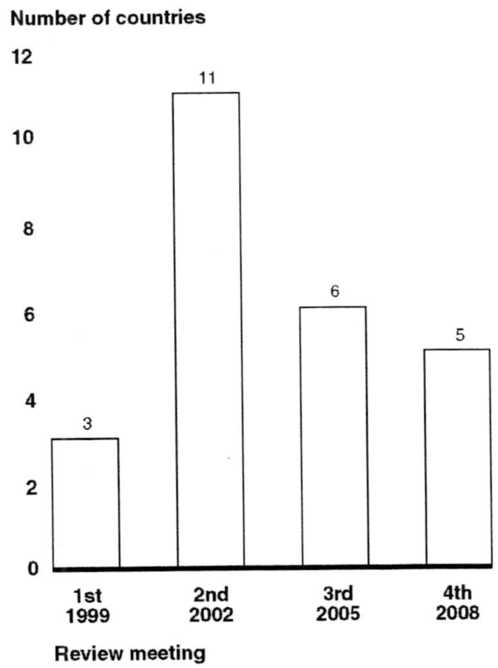

Source: GAO analysis based on IAEA data.

Figure 2. Number of Countries Posting Responses to Questions Received on Their National Reports on IAEA's Public Web Site, 1999-2008

Universal Participation Would Advance Achievement of the Convention's Goals

The Convention is designed to maximize the number of countries that will participate in order to achieve its goal of promoting the safe operation of civilian nuclear power reactors worldwide; however, it is voluntary in nature. By and large, this approach has worked. Since 2009, three countries that are considering developing civilian nuclear power programs—Libya, Jordan, and the United Arab Emirates—have become parties to the Convention. Two others—Kazakhstan and Saudi Arabia—approved the Convention in 2010 and are expected to become parties to it later this year. An overwhelming majority of the parties we surveyed and interviewed said that all countries should be encouraged to join as soon as possible after making the decision to consider developing a nuclear power program. At present, all countries with such programs—except Iran—are parties to the Convention. Several parties we interviewed told us that Iran, which is on the verge of commissioning civilian nuclear power reactors, should ratify the Convention in order to benefit from the safety expertise that participation provides. In their view, without Iran's participation in the Convention, the international community has limited or no insight on, or access to, how Iran is developing, operating, and maintaining its burgeoning civilian nuclear power program. Russian officials with whom we spoke agreed that greater international access to Iran's civilian nuclear power program is needed and that the Convention could play a role in providing that access. Russia is helping Iran build the civilian nuclear power reactor at Bushehr, which is expected to be commissioned in the near future. Russian Ministry of Foreign Affairs officials told us that Russia's continued assistance to Iran's civilian nuclear program may be conditioned on Iran's becoming a party to the Convention.

The Convention is Not a Mechanism for Shutting down Unsafe Reactors

The Convention does not require that unsafe reactors be closed down. As noted in our 1999 testimony,[18] the Convention neither provides sanctions for noncompliance with any of its safety obligations nor does it require the closing of any unsafe nuclear reactors. However, more than 13 years after the Convention came into force, Russia continues to operate 11 Chernobyl-style RBMK reactors.[19] These reactors pose the highest risk, according to Western safety experts, because of their inherent design deficiencies, including their lack of a containment structure. The containment structure, generally a steel-lined concrete dome, serves as the ultimate barrier to the release of radioactive material in the event of a severe accident. Russian nuclear regulators told us that adequate safety upgrades have been made to all 11 RBMK reactors and that they will continue to operate for the foreseeable future. We also discussed the matter of shutdown of Soviet-designed reactors with EU officials, who told us that the Convention was never intended to be a mechanism for closing unsafe Soviet-designed reactors. The European Union has used a different strategy to accomplish shutdown of the unsafe nuclear reactors in its member countries: making EU membership contingent upon the closure of these reactors. As a result, all eight RBMK and first-generation VVER 440 Model 230 reactors in Bulgaria, Lithuania, and Slovakia have been permanently shut down in order for these countries to obtain EU membership.[20]

According to NRC officials, as is the case in other international law on reactor safety, under the Convention each country is responsible for regulating the safety of its own reactors.

In addition, NRC noted that the Convention relies on the peer review process, that it cannot obligate countries to comply with safety standards, and that it does not provide for sanctions such as the closing of any unsafe nuclear power plants. State expressed a similar view. State pointed out that the Convention was never meant to have the authority to require that unsafe reactors be shut down. According to State, it is the position of IAEA and its member states that each country operating nuclear power plants should have its own nuclear regulatory agency that would have the authority to shut down plants.

STEPS HAVE BEEN TAKEN TO IMPROVE THE CONVENTION'S PEER REVIEW PROCESS, AND ADDITIONAL PROPOSALS ARE BEING CONSIDERED

The parties to the Convention generally agree that it would be difficult to amend the Convention. Consequently, several parties have taken the lead in making changes to the Convention's rules and procedures. To date, some steps have been taken to improve the Convention's peer review process, and parties are considering several additional proposals.

Changes Have Been Adopted to Improve the Peer Review Process

Several parties have focused on improving the workings of the Convention's peer review process. The most significant change they have made, in our view, is to allow the parties to more freely ask questions about each others' national reports. NRC expressed concern in our January 1997 report about the rules governing how parties' country group assignments affect the parties' ability to discuss and seek clarification about other parties' national reports at review meetings.[21] According to NRC officials, in the past, parties assigned to a particular country group could ask questions about other parties' nuclear programs that were assigned to that group during the question-and-answer session following the presentation of a national report. However, parties that were not assigned to that country group could not ask questions unless they submitted a written question several months in advance of the review meeting. This restrictive practice began to change during the 2005 review meeting, when at least one country group allowed parties that were not assigned to it to ask questions. At the next review meeting in 2008, according to NRC officials who attended both meetings, no restrictions were placed on any parties' ability to ask questions about the national reports of any other parties. An NRC official told us that this change has made the process more open and accessible to all of the parties.

Another notable change to the rules and procedures of the peer review process is the recent decision to move up the date for the organizational meeting and the selection of officers for the upcoming review meeting by almost a year and to advance by a few weeks the deadlines for submitting national reports and written questions for the peer review process. The purposes of the organizational meeting, among other things, are to elect the officers for the upcoming review meeting,[22] adopt a provisional agenda for the meeting, assign parties to particular country groups, and identify which proposals for enhancing the peer review process should be considered at the upcoming meeting. Previously, organizational meetings were held about 7 months before the upcoming review meeting. However, the parties at the 2008 review meeting agreed to hold the organizational meeting for the 2011 review meeting in September 2009—19 months in advance. According to NRC officials, the purpose of the scheduling change was to put officers in place earlier to give them more time to plan for the next meeting and to promote greater continuity from one meeting to the next. Moving up the deadlines for submitting national reports and written questions for peer review is intended to give countries more time to both review the national reports of other parties and answer any written questions submitted.

Parties are Considering Additional Proposals to Improve the Implementation of the Convention

Additional proposals to improve the implementation of the Convention are currently under consideration by the parties. Specifically, these proposals include (1) allocating more country group meeting time to discuss, among other things, the national reports of countries with emerging nuclear programs; (2) expediting the process for calling a special meeting between review meetings to discuss urgent safety issues; and (3) changing the process for assigning parties to country groups.

Allocating More Time to Countries with Emerging Nuclear Programs

Some parties have suggested the peer review process might be more effective if more review meeting time were allocated to discussing the national reports of countries with emerging nuclear power programs or topics of general concern and less time presenting and discussing the national reports of parties with well-established nuclear programs. For example, according to NRC officials, the United Arab Emirates, which has only recently become a party to the Convention, is rapidly moving to establish its nuclear regulatory infrastructure and is soon to begin construction of several nuclear power reactors. Because its civilian nuclear power program is so new, the United Arab Emirates could benefit from more time to present its national report during the peer review process. NRC officials told us that the United States, in contrast, does not need as much time as it is allocated to present its national report. Similarly, according to a senior NRC official, the United States has proposed that more time at review meetings might also be allocated to discuss topics of general concern—such as the safety challenges of dealing with aging reactors or the challenges parties face in maintaining adequate staffing and competence levels in both the regulatory bodies and at nuclear power plants.

Expediting the Process for Calling Special Meetings

Another proposal to be considered would create a more efficient process for calling a meeting to discuss topical or urgent nuclear safety issues that parties feel cannot wait until the next review meeting. Currently, in order to have such a meeting, a majority of parties are required to support the call for a meeting. One way of streamlining this process, according to an NRC official, would be to empower the officers elected for the most recent or upcoming review meeting to call a special meeting. An urgent issue might be, for example, a nuclear power plant accident. If such an accident occurred, parties might wish to convene a special meeting to discuss the causes of the accident and what might be done to avoid a similar accident.

Changing the Process for Assigning Parties to Country Groups

Finally, to promote greater variation in the composition of country groups from meeting to meeting, amending the method for assigning countries to the six country groups is being considered.[23] Specifically, the experience of the first four review meetings has been that the country groups have remained relatively static—that is, there has been little variation in the membership of each group among the nuclear power countries. According to NRC officials, it would be useful if the composition of the groups were more varied from meeting to meeting. While each group would still be anchored by a country with a large number of operating civilian nuclear power reactors, the remainder of the group would consist of a more varied mix of countries. This type of mix would provide greater opportunities for more information sharing among a more diverse group of countries. An NRC official told us that many parties are generally in favor of some adjustment to the existing process but that there is not yet sufficient agreement on how to accomplish this change.

IAEA's Assistance Programs to Member States Complement the Convention's Safety Goals and Objectives

IAEA has a long history of serving as a technical advisor to member states to promote the safe operation of nuclear power plants. Although this role predates the establishment of the Convention, and regulating nuclear safety is a national responsibility, the Convention complements the role the agency plays in these matters. IAEA promotes the Convention's nuclear safety goals and objectives largely through its Technical Cooperation (TC) Program, safety standards, and peer review missions, which together help countries improve their nuclear regulatory bodies and the safety performance of their civilian nuclear power plants. Most survey respondents reported that they found IAEA effective in serving as a technical advisor. In addition, almost all parties responding to our survey consider IAEA to be effective in its role as secretariat to the Convention.

IAEA's Technical Cooperation Program, Safety Standards, and Peer Review Missions Play an Important and Growing Role in Promoting Nuclear Safety Worldwide

IAEA provides assistance to its member states to promote peaceful uses of nuclear energy in several ways, including providing technical cooperation, establishing safety standards, and conducting advisory and peer review missions. The importance of its role in providing this type of assistance was corroborated by our survey results. A majority of survey respondents reported that IAEA was either very effective or somewhat effective in serving as a technical advisor to countries requesting assistance to improve civilian nuclear power safety. IAEA's TC program supports, among other things, nuclear safety and the development of nuclear power.[24] For the 2009-2011 activities under the TC program, nuclear safety remains one of the top three priorities for IAEA member states. IAEA currently conducts 551 TC projects in 115 countries and territories, and program activities are tailored to the needs of each region. Specific TC projects have included activities to extend the operating life of nuclear power plants and establishing safety culture in nuclear facilities. TC projects that support member states considering or developing nuclear power also include strengthening nuclear regulatory authorities and preparing an emergency plan for a nuclear power plant. In 2007, IAEA disbursed approximately $5.6 million to support the safety of civilian nuclear installations worldwide through the TC program. In addition to its TC program budget, IAEA plans to spend approximately $15.1 million in 2010 on other efforts to promote nuclear safety, such as strengthening countries' abilities to respond to nuclear incidents and emergencies and to assess the safety of the siting and design of nuclear installations. The role and importance of IAEA in promoting nuclear safety will likely grow if the cost of fossil fuels and the threat of climate change spur a nuclear renaissance, as an independent commission assessing the role of IAEA to 2020 and beyond reported recently.[25] According to this independent commission, this growing role may involve (1) leading an international effort to establish a global nuclear safety network, (2) helping countries with emerging nuclear power programs put in place the infrastructure needed to develop nuclear energy safely, and (3) ensuring that critical safety knowledge is widely shared among IAEA member states.

In addition, IAEA has established safety standards that provide a framework for fundamental safety principles, requirements, and guidance for member states. The standards, which reflect international consensus, cover a wide range of topics, including nuclear power plant design and operation, site evaluation, and emergency preparedness and response. Committees of senior experts from IAEA member states use an open and transparent process to develop the standards and any subsequent revisions. The guidelines governing the drafting of national reports state that IAEA safety standards can give valuable guidance on how to meet the Convention's safety obligations.

IAEA also promotes nuclear safety through advisory and voluntary peer review missions—the most prominent are Integrated Regulatory Review Service (IRRS) missions and Operational Safety Review Team (OSART) missions. These missions evaluate the operations of a member state's nuclear regulatory system and civilian nuclear power plant operational safety, respectively. IRRS missions assess the safety practices of the requesting country through an examination of its regulatory framework and organization and compare the country's practices with IAEA safety standards. Since 1992, IAEA has conducted 44 IRRS missions in 26 countries, with 15 of these missions taking place in countries that have operated—and in some cases continue to operate—Soviet-designed reactors. Table 1 shows the number of IRRS missions that member countries had hosted through 2009. The United States has sent approximately 20 experts on IRRS missions and has agreed to host an IRRS mission in October 2010.

Some parties that responded to our survey reported that they found IRRS and OSART missions effective at improving civilian nuclear power safety. In addition, according to the summary report of the Convention's fourth meeting in 2008, many parties reported that they had positive experiences with IRRS and OSART missions, and parties who had not already hosted one of these missions were encouraged to do so. In February and March 2010, IAEA conducted an IRRS mission to Iran, which included a site visit to the nearly completed Bushehr nuclear power plant. IAEA recommended, among other things, that Iran join the Convention.

Table 1. Number of IRRS Missions by Country, 1992 through 2009

Country	Number of IRRS missions[a]
Armenia	2
Australia	1
Bulgaria	2
Canada	1
China	2
Czech Republic	2
Finland	2
France	2
Germany	1
Hungary	2
Indonesia	1
Japan	2
Lithuania	1
Malaysia	2
Mexico	2
Pakistan	1
Peru	1
Romania	4
Russia	1
Slovakia	2
Slovenia	1
Spain	1
Switzerland	2
Ukraine	3
United Kingdom	2
Vietnam	1
Total	**44**

Source: GAO analysis of IAEA data.

[a] IRRS missions were preceded by a similar program from 1992-2004 called International Regulatory Review Team missions. This table combines numbers for both types of missions.

According to a senior Swedish official who was involved in drafting the Convention, these missions are increasingly being used to measure the safety standards of parties to the Convention. Parties face peer pressure to submit to these voluntary missions, as they provide a way for a country to show its commitment to enhancing safety. For example, ENSREG has promoted the use of IRRS missions by EU countries. Describing the missions as "well established and well respected," ENSREG has encouraged all EU member states to participate in one to obtain advice on improvements and to learn from the best practices of others.

IAEA also manages the OSART missions through which teams of experts drawn from IAEA member countries—including the United States, which has sent over 100 experts on missions—review operational safety at specific nuclear power plants. IAEA has conducted over 150 OSART missions in 32 countries since 1983, and has 9 more scheduled through the end of 2011. Table 2 shows the number of OSART missions that member countries had hosted through 2009.

As table 2 shows, the 2 countries that have hosted the most OSART missions are France and Ukraine, 21 and 14, respectively. Combined, those 2 countries have 73 reactors. China and the Czech Republic have hosted the second most missions, 10 and 8, respectively. These countries have a combined total of 17 operating reactors. Japan, which has 54 reactors, has hosted 5 OSART missions. Russia, which has 32 operating reactors, has hosted 6, and the United States, which has 104 operating reactors, has also hosted 6 missions. The only countries with operating civilian nuclear power programs that have not hosted OSART missions are Armenia and India, which operate 1 and 18 reactors, respectively.

Table 2. Number of OSART Missions by Country, 1983 through 2009

Country	Number of OSART missions
Argentina	1
Belgium	1
Brazil	5
Bulgaria	6
Canada	3
China	10
Czech Republic	8
Finland	3
France	21
Germany	6
Hungary	2
Italy	2
Japan	5
Kazakhstan	1
Korea, Republic of	6
Lithuania	2
Mexico	4
Netherlands	3
Pakistan	5
Philippines	2
Poland	1
Romania	3
Russia	6
Slovakia	5
Slovenia	3
South Africa	3
Spain	5
Sweden	6
Switzerland	4
Ukraine	14
United Kingdom	3
United States	6
Total	**155**

Source: GAO analysis of IAEA data.

While recommendations that result from safety review services such as IRRS and OSART missions are not mandates, IAEA officials told us that the agency nevertheless sees a high rate of implementation of those recommendations. IAEA also makes available on its public Web site a compilation of best practices learned from recent OSART missions, as well as the mission reports as authorized by the member states. This compilation serves to help member states improve the operational safety of their power plants and includes emergency plans and preparedness, training, and maintenance.

Finally, IAEA also promotes civilian nuclear safety through other means. For example, IAEA offers additional review services to member states by focusing on issues such as siting, seismic safety, research reactor safety, fuel cycle facilities' safety, power plant accident management, and safety culture assessments. IAEA also promotes education and training in nuclear safety through Web-based courses, electronic textbooks, and workshops. This training covers topics such as basic safety concepts, regulatory control of nuclear power plants, and instruction on IAEA safety standards. Much of this information is available to the public to download from IAEA's Web site. One survey respondent from Eastern Europe commented that the training courses and workshops had contributed significantly to the promotion of high safety standards and best practices. Moreover, IAEA regularly holds conferences and symposia on issues related to nuclear safety, with some event summaries available online. Recent topics have included promoting safety education and training for countries with new or expanding nuclear programs, ensuring safety for sustainable nuclear development, and managing nuclear power plant life.

IAEA Is Effective as the Convention's Secretariat, according to almost All Survey Respondents and Parties We Interviewed

Almost all parties responding to our survey and parties we interviewed reported that IAEA effectively carries out its role as secretariat as outlined in the Convention. In this capacity, IAEA hosts the review meetings in Vienna, Austria; prepares documents; and provides translation and interpretation services. There was widespread agreement among the respondents that the agency is effective in convening, preparing, and servicing the meetings and at transmitting information received or prepared in accordance with the provisions of the Convention.

Some survey respondents and parties we interviewed called for more IAEA support during the Convention's review meetings in such areas as more translation services for all country group sessions and more administrative assistance for parties to the Convention. The Convention permits IAEA to provide other services in support of the review meetings, if the parties reach consensus. Finally, some survey respondents reported that IAEA should play a more active role in the following areas:

- helping prepare national reports,
- providing other assistance to help prepare for the next review meeting,
- providing other technical support to improve safety, and
- helping address concerns about a country's civilian nuclear power program.

IAEA estimates its costs to support the last review meeting in 2008 at nearly $118,000 and expects to spend approximately $130,000 for the fifth review meeting scheduled for April 2011. The costs associated with the review meetings are modest for the U.S. government as well. NRC and State spent approximately $725,000 preparing for and participating in the 2008 review meeting and estimate they will spend $825,000 for the next review meeting.

CONCLUSIONS

The Convention plays an important role in strengthening nuclear safety and enjoys broad support among the parties we surveyed and interviewed. Support for the Convention continues to grow as evidenced by the increasing number of countries that have joined it, particularly those with emerging nuclear programs, such as the United Arab Emirates. Many parties to the Convention told us that all countries that are considering embarking on a nuclear power program—or currently operating civilian nuclear power reactors— should be encouraged to join the Convention, including Iran.

We are encouraged that the parties have taken steps to improve the Convention's peer review process. However, the Convention does not require parties to include performance metrics in their national reports, which makes it difficult to gauge its impact on improving nuclear safety. Without such metrics there is no systematic way to measure where and how progress has been made in improving safety in each country that operates civilian nuclear power reactors. In addition, more than half of the survey respondents reported that the lack of metrics hampers the Convention's usefulness, and NRC has noted that it would be feasible to add performance metrics into the guidelines that implement that national report process called for by the Convention. There are already international organizations that use such indicators to track nuclear safety improvements and which could perhaps be incorporated into the guidelines as voluntary practices that parties are encouraged to implement. Further, public awareness about parties' progress toward meeting the terms of the Convention can play a key role in influencing compliance with the Convention's nuclear safety obligations. However, to date the public has had limited access to parties' national reports and written answers to questions about their nuclear power programs. More than half of the national reports prepared for the 2008 review meeting are not posted to IAEA's public Web site, and even fewer parties make their answers to written questions received on their national reports available on IAEA's public Web site. Putting this information on the Web site could increase public awareness of the nuclear safety issues facing countries and how they are addressing them.

RECOMMENDATIONS FOR EXECUTIVE ACTION

To further enhance the usefulness of the Convention in promoting the safety of civilian nuclear power programs worldwide, we recommend that the Secretary of State, in coordination with the Chairman of the Nuclear Regulatory Commission, work with other parties to the Convention to take the following three actions:

- Encourage parties to include performance metrics in national reports to better track safety in civilian nuclear power plants and help countries more systematically measure where and how they have made progress in improving safety.
- Expand efforts to increase the number of parties' national reports made available to the public by posting them to IAEA's public Web site.
- Promote greater public dissemination of parties' written answers to questions about their nuclear power programs by posting this information to IAEA's public Web site.

AGENCY COMMENTS AND OUR EVALUATION

We provided a draft of this chapter to NRC and State for comment. We also provided IAEA with a detailed summary of facts contained in the draft report. State and NRC provided written comments on the draft report, which are presented in appendixes IV and V, respectively. IAEA, State, and NRC also provided technical comments, which we incorporated as appropriate.

NRC generally agreed with our report but did not specifically agree or disagree with the report's recommendations, and State generally agreed with the recommendations to (1) encourage parties to the Convention to include performance metrics in their national reports to better track safety in civilian nuclear power plants, (2) increase the number of parties' national reports made available to the public by posting them to IAEA's public Web site, and (3) promote greater public dissemination of parties' written answers to questions about their nuclear power programs by posting this information to IAEA's public Web site. In its written comments, however, State provided some clarifications concerning the recommendations. First, State noted that it might be difficult to achieve metrics that would be meaningful across so many countries' nuclear power programs and to agree on the specific metrics to be used. Second, State noted that initiatives to increase public access to information would run counter to strong concerns regarding confidentiality of information on civilian nuclear power plants held by many parties. In addition, State asserted that the report somewhat mischaracterizes the Convention by noting that the Convention does not require that unsafe reactors be shut down. State noted that the Convention was never meant to have that authority, which would be contrary to IAEA practice and policy. It is the position of IAEA and member states that each country operating nuclear power plants should have its own national regulatory agency that would have the authority to shut down plants.

Regarding the first point, while it might be challenging to establish a common set of performance metrics, we believe there are already examples of standard metrics being used, such as those published by WANO. We believe that WANO's metrics, for instance, could be used as a benchmark for parties to follow in measuring safety progress when developing their national reports. With regard to encouraging public dissemination of information about the Convention, we agree that maintaining confidentiality of sensitive information about what is discussed among the parties during the peer review process should be maintained. However, we also believe that increasing public awareness of the Convention's proceedings—even on an incremental basis—through the posting of national reports to IAEA's public Web site is a worthwhile goal and should be encouraged to the extent practicable.

Finally, with respect to the issue of unsafe reactors, we have not mischaracterized the Convention. Rather, we pointed out in the report—as we have previously reported—that the Convention does not require the closing of any unsafe nuclear reactors. We also noted in this chapter that nuclear safety is a national responsibility and have not suggested or implied that the Convention is flawed because it does not require unsafe reactors to be closed. The fact remains, however, that Russia, which has ratified the Convention, continues to operate numerous nuclear power plants that pose a safety risk according to Western safety experts. However, based on State's comments, we have clarified the text regarding this issue.

APPENDIX I: PARTIES TO THE CONVENTION ON NUCLEAR SAFETY

Country	Number of civilian nuclear reactors	Entry into force
Argentina[a]	2	16 July 1997
Armenia[a]	1	20 December 1998
Australia	0	24 March 1997
Austria	0	24 November 1997
Bangladesh	0	24 October 1996
Belarus	0	27 January 1999
Belgium[a]	7	13 April 1997
Brazil[a]	2	2 June 1997
Bulgaria[a]	2	24 October 1996
Canada[a]	18	24 October 1996
Chile	0	20 March 1997
China[a]	11	24 October 1996
Croatia	0	24 October 1996
Cyprus	0	15 June 1999
Czech Republic[a]	6	24 October 1996
Denmark	0	11 February 1999
Estonia	0	4 May 2006
Finland[a]	4	24 October 1996
France[a]	58	24 October 1996
Germany[a]	17	20 April 1997
Greece	0	18 September 1997
Hungary[a]	4	24 October 1996
Iceland	0	2 September 2008
India[a]	18	29 June 2005
Indonesia	0	11 July 2002
Ireland	0	24 October 1996
Italy	0	14 July 1998
Japan[a]	54	24 October 1996
Jordan	0	10 September 2009
Korea, Republic of[a]	20	24 October 1996
Kuwait	0	9 August 2006

(Continued)

Country	Number of civilian nuclear reactors	Entry into force
Latvia	0	23 January 1997
Lebanon	0	24 October 1996
Libya	0	11 November 2009
Lithuania	0	24 October 1996
Luxembourg	0	6 July 1997
Mali	0	24 October 1996
Malta	0	13 February 2008
Mexico[a]	2	24 October 1996
Netherlands[a]	1	13 January 1997
Nigeria	0	3 July 2007
Norway	0	24 October 1996
Pakistan[a]	2	29 December 1997
Peru	0	29 September 1997
Poland	0	24 October 1996
Portugal	0	18 August 1998
Republic of Moldova	0	5 August 1998
Romania[a]	2	24 October 1996
Russian Federation[a]	32	24 October 1996
Saudi Arabia	0	16 June 2010[b]
Senegal	0	24 March 2009
Singapore	0	15 March 1998
Slovakia[a]	4	24 October 1996
Slovenia[a]	1	18 February 1997
South Africa[a]	2	24 March 1997
Spain[a]	8	24 October 1996
Sri Lanka	0	9 November 1999
Sweden[a]	10	24 October 1996
Switzerland[a]	5	11 December 1996
The FYR of Macedonia	0	13 June 2006
Turkey	0	24 October 1996
Ukraine[a]	15	7 July 1998
United Arab Emirates	0	29 October 2009
United Kingdom[a]	19	24 October 1996
United States[a]	104	10 July 1999
Uruguay	0	2 December 2003
EURATOM	0	30 April 2000
Total	**437**	

Source: IAEA.

Note: The total of 437 reactors represents the reactors in the list plus 6 reactors in Taiwan, which IAEA includes in the total number worldwide.

[a] Indicates that the state has at least one nuclear installation that has achieved criticality in a reactor core.

[b] Anticipated date of entry into force. Saudi Arabia deposited its instrument of accession to the Convention on March 18, 2010. By the terms of the Convention, it will enter into force for Saudi Arabia 90 days after the date of deposit of the instrument of accession.

APPENDIX II: INFORMATION ON U.S. AND EUROPEAN UNION FUNDING TO PROMOTE INTERNATIONAL NUCLEAR SAFETY

United States

Table 3 reflects the cumulative amount of nuclear reactor safety assistance funds provided by the Department of Energy (DOE) from the inception of these programs.

Table 4 reflects the cumulative amount of nuclear reactor safety assistance funds provided by the Nuclear Regulatory Commission (NRC) from the inception of these programs.

Table 3. Obligations and Expenditures for DOE's Safety Assistance Programs as of September 30, 2009

Dollars in thousands					
Recipient	**Funds available**	**Funds unobligated**	**Funds obligated**	**Funds obligated and spent**	**Funds obligated but not spent**
Ukraine	$369,223	$0	$369,223	$360,918	$8,305
Russia	179,917	0	179,917	179,917	0
Armenia	50,813	0	50,813	47,734	3,079
Kazakhstan	7,732	0	7,732	7,317	415
Noncountry specific	73,269	0	73,269	73,269	0
DOE subtotal	$725,634	$0	$725,634	$713,659	$11,975

Source: DOE.

Notes:

Expenditures identified in this table are not linked to the Convention on Nuclear Safety. Rather, they refer only to DOE bilateral assistance programs to support nuclear safety efforts in various foreign countries.

According to DOE, funding appropriation end dates for the programs are as follows:

Ukraine: 2008 Russia: 2005

Central and Eastern Europe: 2006

Armenia: 2011 (estimated) Kazakhstan: 2007 Noncountry specific: 2004

According to DOE, all programs will expend funds through at least fiscal year 2010, with the exception of Russia, which ceased expending funds in fiscal year 2006.

Table 4. Obligations and Expenditures for NRC's Reactor Safety Assistance Programs as of September 30, 2009

Dollars in thousands					
Recipient	**Funds available**	**Funds unobligated**	**Funds obligated**	**Funds obligated and spent**	**Funds obligated but not spent**
Ukraine	$22,083	$0	$22,083	$21,482	$601
Russia	17,794	0	17,794	17,493	301
Central and Eastern Europe	8,044	0	8,044	8,044	0
Armenia	7,715	0	7,715	6,899	816

Table 4. (Continued)

Dollars in thousands					
Recipient	Funds available	Funds unobligated	Funds obligated	Funds obligated and spent	Funds obligated but not spent
Kazakhstan	6,920	0	6,920	6,920	0
Total	**$62,556**	**$0**	**$62,556**	**$60,838**	**$1,718**

Source: NRC.

Notes:

According to NRC, these funds are provided through the Support for Eastern European Democracies (SEED) Act, which funded Central and Eastern European countries, and through the Freedom Support Act (FSA), which funds Armenia, Kazakhstan, Russia, and Ukraine. SEED Act figures are cumulative from fiscal year 1991, and FSA figures are cumulative from fiscal year 1992.

These expenditures identified in this table are not linked to the Convention on Nuclear Safety. Rather, they refer only to NRC bilateral assistance programs to support nuclear safety efforts in various foreign countries.

According to NRC, fiscal year 2008 was the last year for which NRC obligated FSA funds for Russia, shifting its focus with Russia to cooperation instead of assistance. NRC will expend all remaining FSA funds for assistance for Russia during fiscal year 2010.

Table 5. Total Nuclear Safety Budget for the Technical Assistance to the Commonwealth of Independent States Program

Dollars in millions	
Year	Amount
1991	$97.8
1992	112.9
1993	145.8
1994	149.8
1995	169.8
1996	198.6
1997	100.9
1998	121.7
1999	86.8
2000	61.2
2001	100.2
2002	86.6
2003	136.5
2004	145.3
2005	96.3
2006	109.4
Total	$1,919.5

Source: "International Nuclear Safety Actions of the European Commission," EuropeAid Co-operation Office. Notes: Figures are in millions of 2010 dollars.

The expenditures identified in this table are estimates and are not linked to the Convention on Nuclear Safety.

The figures include funding for the Russian Federation, the Northern Dimension Fund to the 'Nuclear Window,' Ukraine, other countries, and the Chernobyl Shelter Fund.

The Technical Assistance to the Commonwealth of Independent States (TACIS) program was replaced in 2007 by the Nuclear Safety Cooperation Instrument (NSCI), which finances measures to support nuclear safety, radiation protection, and safeguards of nuclear materials. The NSCI has a budget of roughly $730 million for 2007-2013.

European Union

Table 5 reflects nuclear safety expenditures from the European Union's Technical Assistance to the Commonwealth of Independent States program.

APPENDIX III: SCOPE AND METHODOLOGY

The objectives of our review were to evaluate the extent to which the Convention on Nuclear Safety is achieving its primary goal: promoting the safe operation of civilian nuclear power reactors worldwide. Specifically, we assessed (1) parties' views on the perceived benefits and limitations of the Convention; (2) efforts to improve the implementation of the Convention; and (3) how International Atomic Energy Agency (IAEA) programs complement the Convention's safety goals and objectives. In addition, we are providing information in appendix II about funding provided by the United States and the EU to promote international nuclear safety since the early 1990s.

To assess parties' views of the perceived benefits and limitations of the Convention and efforts to improve implementation, we (1) interviewed representatives of 17 nuclear and nonnuclear parties to the Convention as well as officials from NRC and State responsible for representing the United States at the Convention; (2) analyzed various Convention-related documents from NRC, State, IAEA, and EU; and (3) conducted a Web-based survey of 64[26] parties to the Convention. To encourage honest and open responses to our survey, we pledged member countries confidentiality[27] in their responses and indicated that we would report only aggregate information or examples that would not identify a particular party. The survey included questions about the usefulness of the Convention, the effectiveness of Convention activities, and the role of IAEA in the Convention.

To develop the survey questions, we analyzed the text of the Convention itself, as well as related rules and procedures. We also interviewed parties to the Convention and other experts to identify issues related to the Convention. Finally, we reviewed previous GAO reports to identify past issues and concerns related to the Convention and developed survey questions to gauge whether these issues were still relevant. The survey was pretested to ensure that (1) the questions were clear and unambiguous, especially to nonnative English-speaking respondents; (2) the terms we used were precise; (3) the survey did not place an undue burden on the officials completing it; and (4) the survey was independent and unbiased. In addition, the survey was reviewed by an independent, internal survey expert and by NRC.

The survey was conducted using self-administered electronic questionnaires posted on the World Wide Web. We sent e-mail notifications to 64 parties to the Convention to alert them that we were conducting the survey and would be sending them log-in information in a separate e-mail. We also e-mailed each potential respondent a unique password and username to ensure that only members of the target population could participate in the survey. To encourage respondents to complete the survey, we sent an e-mail reminder to each nonrespondent about 2 weeks after our initial e-mail message. We also sent an additional e-mail reminder that extended the deadline to complete the survey. In addition to these e-mails, we also conducted extensive telephone and personalized e-mail follow-up to encourage those parties who contacted us with questions about the survey and to encourage the nonrespondents from the 17 parties whose representatives we interviewed to complete the survey. The survey data were collected from October 2009 through December 2009. Half (32) of the 64 parties to the Convention responded to the survey. To assess the potential for nonresponse bias in our survey results, we compared selected characteristics of nonresponding countries, such as (1) length of time as a party to the Convention, (2) nuclear power status and number of nuclear power plants, (3) region, (4) former Soviet bloc alignment, and (5) EU membership, to those of the responding parties. The distribution of these characteristics among responding and nonresponding parties was well-balanced. For example, 3 of the 32 respondents have been parties to the Convention for 2 years or less, 2 respondents for 3 to 9 years, and 27 respondents for 10 or more years. In addition, we also received responses from 13 nonnuclear countries and 19 nuclear countries and 17 EU-member countries and 15 nonmember countries. To eliminate data-processing errors, we independently verified the computer program that generated the survey results. This chapter does not contain all the results from the survey; the survey and a more complete tabulation of the results are provided in an electronic supplement to this chapter (this supplement can be viewed online at GAO-10-550SP).

To assess how IAEA programs complement the Convention's safety goals and objectives, we analyzed budget and other relevant documents from the Convention, such as meeting minutes and rules of procedure. We also interviewed IAEA officials; U.S. officials at the U.S. Missions in Vienna and Brussels; and the representatives of 17 parties to the Convention in Vienna, Brussels, Moscow, and Washington, D.C. To determine the amount of money the United States has spent promoting nuclear safety from the early 1990s through September 30, 2009, we obtained expenditure information from DOE and NRC. To assess the reliability of the information provided, we interviewed knowledgeable officials from each agency to understand (1) how they had developed the estimates and (2) what supporting documentation had been used to develop them; we determined the information provided was sufficiently reliable for our purposes. To determine the amount of money the EU has spent promoting nuclear safety from 1991 through 2006, and the amount they have budgeted to spend from 2007 to 2013, we obtained budget information from EU officials. However, the reliability of these EU estimates is undetermined because we did not receive responses to our data reliability questions. Given these limitations, we characterize these costs as estimates, and we use them only as background. Because the EU budget information was provided in euros, we converted the original values to dollars. In all instances, when converting euros to dollars, we used nominal and purchasing power parity average annual exchange rates from the Organization for Economic Cooperation and Development. When converting euro values for future projections into dollars, we used the latest available annual exchange rate. In addition, to determine the amount of money IAEA has budgeted for nuclear safety in 2010, we obtained information from the agency's *Programme and Budget for 2010-11*. These IAEA budget figures—which we converted to dollars from euros—are also of undetermined reliability because we were unable to obtain sufficient detail about how they developed the estimates or the data sources that supported them. To determine the cost to the United States to participate in the Convention, and IAEA's costs to support the Convention for one 3-year cycle, we obtained expenditure information from NRC, State, and IAEA. To assess the reliability of this information, we also interviewed knowledgeable officials from each agency to understand (1) how they had developed the estimates and (2) what supporting documentation had been used to develop them. We determined the information provided by NRC was sufficiently reliable for our purposes. However, the reliability of the State and IAEA information is undetermined. The reliability of State estimates are unknown because staff typically combined work and travel related to the Convention with other work duties, so it is not possible to accurately determine the amount of money spent exclusively on Convention participation. IAEA estimates—which we converted to dollars from euros—are of undetermined reliability because they do not formally track costs to run the review meetings.

We conducted this performance audit from February 2009 to April 2010, in accordance with generally accepted government auditing standards. Those standards require that we plan and perform the audit to obtain sufficient, appropriate evidence to provide a reasonable basis for our findings and conclusions based on our audit objectives. We believe that the evidence obtained provides a reasonable basis for our findings and conclusions based on our audit objectives.

APPENDIX IV: COMMENTS FROM THE DEPARTMENT OF STATE

United States Department of State

Chief Financial Officer

Washington, D.C. 20520

APR 19 2010

Ms. Jacquelyn Williams-Bridgers
Managing Director
International Affairs and Trade
Government Accountability Office
441 G Street, N.W.
Washington, D.C. 20548-0001

Dear Ms. Williams-Bridgers

We appreciate the opportunity to review your draft report, "NUCLEAR SAFETY: Convention on Nuclear Safety Is Viewed by Most Member Countries as Strengthening Safety Worldwide," GAO Job Code 361054.

The enclosed Department of State comments are provided for incorporation with this letter as an appendix to the final report.

If you have any questions concerning this response, please contact Jan Fladeboe, Foreign Affairs Officer, Bureau of International Security and Nonproliferation at (202) 647-6957.

Sincerely,

James L. Millette

cc: GAO – Glen Levis
ISN – Vann Van Diepen
State/OIG – Tracy Burnett

Department of State Comments on GAO Draft Report

NUCLEAR SAFETY:
Convention on Nuclear Safety Is Viewed by Most Member Countries as Strengthening Safety Worldwide
(GAO-10-489, GAO Code 361054)

Thank you for the opportunity to comment on your draft report entitled "NUCLEAR SAFETY: Convention on Nuclear Safety Is Viewed by Most Member Countries as Strengthening Safety Worldwide." With the anticipated growth in the number of nations worldwide that operate nuclear power plants, the role of the Convention on Nuclear Safety will become more important in ensuring the safe operation of these plants.

Promotion of the safe operation of nuclear reactors worldwide is one of the U.S. Government's top foreign policy and national security priorities. The Convention on Nuclear Safety plays a very important role in the realization of that. However, it must be understood that the Convention is an incentive instrument. It is not designed to ensure fulfillment of obligations by Parties through control and sanction but is based on their common interest to achieve higher levels of safety which will be developed and promoted through regular meetings of the Parties. The Convention obliges Parties to submit reports on the implementation of their obligations for "peer review" at meetings of the Parties to be held at the IAEA. The GAO report focuses on suggested changes to these reports and their promulgation to the public.

The Department of State generally concurs with the recommendation to encourage Parties to include performance metrics in national reports to better track safety in civilian nuclear power plants and help countries more systematically measure where and how they have made progress in improving safety. However, it must be recognized that it will be difficult to achieve metrics that would be meaningful across so many countries' nuclear power programs. Further, due to the consensus nature of the Convention, it will also be difficult to agree on the specific metrics to be used.

The Department of State generally concurs in efforts to increase the numbers of Parties' national reports made available to the public by posting them to the IAEA's public Web site. However, this initiative will run counter to strong concerns regarding confidentiality of information on civilian nuclear power plants held by many Parties.

While the Department of State generally concurs in the promotion of greater public dissemination of Parties' written answers to questions about their nuclear power programs by posting this information to the IAEA's public Web site, again the concern of Parties over the confidentiality of information on their respective civilian nuclear power programs will make this problematical.

All of the above suggestions are in conflict with the IAEA's precept that it is up to each Party to determine what information should be made public and what should remain confidential. This is an important aspect of the Convention and its implementation, and is one of the factors that have convinced countries to join the Convention.

In addition, the GAO report somewhat mischaracterizes the Convention in its criticism that it "does not require that unsafe reactors should be shut down." The Convention was never meant to have that authority, which would be contrary to IAEA practice and policy. It is the position of the IAEA and Member States that each country operating nuclear power plants should have its own national regulatory agency that would have the authority to shut down plants. If the GAO report's position were taken, then the United States would be in a situation where an outside international body would have the authority to shut down a U.S. reactor, even over the Nuclear Regulatory Commission's authority. Further, the Convention does not require the imposition of sanctions when countries do not follow the safety principles the Parties are called upon to follow.

Appendix V: Comments from the Nuclear Regulatory Commission

UNITED STATES
NUCLEAR REGULATORY COMMISSION
WASHINGTON, D.C. 20555-0001

April 16, 2010

Mr. Gene Aloise, Director
Natural Resources and Environment
Government Accountability Office
441 G Street, NW
Washington, DC 20548

Dear Mr. Aloise:

I would like to thank you for the opportunity to review and submit comments on the April 2010 draft of the U.S. Government Accountability Office (GAO) report, "Convention on Nuclear Safety is Viewed by Most Member Countries as Strengthening Safety Worldwide."

In general, the U.S. Nuclear Regulatory Commission (NRC) agrees with the draft GAO report. However, I am providing certain technical comments concerning the Convention on Nuclear Safety (CNS) in the attached enclosure.

Should you have any questions about these comments, please contact Mr. Jesse Arildsen of my staff at (301) 415-1785 or at Jesse.Arildsen@nrc.gov.

Sincerely,

R. W. Borchardt
Executive Director for Operations

Enclosure:
NRC Technical Comments Regarding GA Draft Report, GAO-10-489

cc: Chairman Jaczko
Commissioner Svinicki
Commissioner Apostolakis
Commissioner Magwood
Commissioner Ostendorff

End Notes

[1] On April 26, 1986, the worst accident in the history of civilian nuclear power occurred at the Chernobyl nuclear power plant in Ukraine, where an explosion destroyed the core of a reactor containing approximately 200 tons of nuclear fuel. The explosion also destroyed much of the reactor building, severed the reactor's cooling pipes, and spewed hot fragments of reactor fuel from the core. The explosion and heat from the reactor core propelled radioactive material up to 6 miles high, where it was then dispersed over 60,000 square miles of land primarily in Ukraine, Belarus, and Russia. Smaller amounts of radioactive material spread over Eastern and Western Europe and Scandinavia and were even detected in the United States.

[2] Safety culture implies individual and organizational awareness of and commitment to the importance of safety. It also refers to the personal dedication and accountability of all individuals engaged in any activity that has a bearing on the safety of nuclear power plants.

[3] Appendix I contains a list of these countries.

[4] The Convention also requires that no more than 3 years pass between meetings held to review national reports.

[5] IAEA, an independent international organization based in Vienna, Austria, that is affiliated with the United Nations, has the dual mission of promoting the peaceful uses of nuclear energy and verifying that nuclear technologies and materials intended for peaceful purposes are not diverted to weapons development efforts. IAEA had 151 member states as of March 2010.

[6] GAO, *Nuclear Safety: Progress Toward International Agreement to Improve Reactor Safety*, GAO/RCED-93-153 (Washington, D.C.: May 14, 1993); GAO, *Nuclear Safety: Uncertainties about the Implementation and Costs of the Nuclear Safety Convention*, GAO/RCED-97-39 (Washington, D.C.: Jan. 2, 1997); and GAO, *Nuclear Safety: The Convention on Nuclear Safety*, GAO/T-RCED-99-127 (Washington, D.C.: Mar. 17, 1999).

[7] At the time we disseminated our survey in October 2009, the Convention had not yet entered into force for two other countries, Libya and the United Arab Emirates, and we did not include them in our survey.

[8] GAO, *Nuclear Safety: Concerns with the Continuing Operation of Soviet-Designed Nuclear Power Reactors*, GAO/RCED-00-97 (Washington, D.C.: Apr. 25, 2000).

[9] GAO, *Nuclear Safety: Construction of the Protective Shelter for the Chernobyl Nuclear Reactor Faces Schedule Delays, Potential Cost Increases, and Technical Uncertainties*, GAO-07-923 (Washington, D.C.: July 19, 2007).

[10] The mission of the NEA is to assist its member countries in maintaining and further developing, through international cooperation, the scientific, technological and legal bases required for the safe and economical use of nuclear energy for peaceful purposes.

[11] WENRA is an organization composed of the chief nuclear regulators of EU countries with nuclear power plants and other interested European countries. WENRA's main objectives are to facilitate the exchange of nuclear safety information and experience among regulators, develop a common approach to nuclear safety, and provide an independent capability to examine nuclear safety in affiliated countries.

[12] ENSREG is an independent, authoritative expert body composed of senior officials from national regulatory or nuclear safety authorities from all 27 member states in the EU.

[13] According to NRC, a critical element of the U.S. international safety assistance administered by NRC since the early 1990s has been to promote the independence and effectiveness of countries' nuclear regulatory authorities.

[14] INPO is a private organization established by American nuclear power plant operators to promote the safe and reliable operation of nuclear power plants.

[15] GAO/T-RCED-99-127.

[16] GAO/T-RCED-99-127.

[17] WANO was established in 1989 to improve nuclear power plant safety worldwide. Every organization in the world that operates a nuclear electricity generating plant is a member of WANO. Members work together to improve nuclear safety through power plant assessments, benchmarks, mutual support, the sharing of information, and the promoting of best practices.

[18] GAO/T-RCED-99-127.

[19] The Soviet-designed RBMK (reactor bolshoy moshchnosty kanalny, or in English, high-power channel reactor) is a pressurized water reactor that uses ordinary water as its coolant and solid graphite (a form of carbon), a very pure form of the same graphite found in pencils, as its moderator. These reactors were favored by the former Soviet Union primarily because, in addition to producing both power (electricity and heat) and plutonium, they were able to be refueled while the reactor was still running. This ability was important to the Soviet Union's national security.

[20] Bulgaria and Slovakia operated a different type of Soviet-designed reactor: the VVER-440-230. The VVER-440-230 is also an inherently unsafe reactor design, according to nuclear safety experts.

[21] GAO/RCED-97-39.

[22] Officers—a president and two vice presidents (one each from a nuclear power country and a nonnuclear country)—are elected for the upcoming review meeting at the organizational meeting. In addition, four

officers are elected for each of the six country groups: a chairperson, vice-chairperson, rapporteur, and coordinator.

[23] Presently, NRC officials told us that parties are assigned to one of six country groups according to their number of operating civilian nuclear power reactors. For example, as the party with the most reactors, the United States is assigned to group 1; France, with the second largest number of reactors, is assigned to group 2; and Japan, with the third largest number, is assigned to group 3. This process continues until all the countries with operating civilian nuclear power reactors are assigned to country groups. Nonnuclear countries are assigned to each of the six groups on a random basis.

[24] GAO, *Nuclear Nonproliferation: Strengthened Oversight Needed to Address Proliferation and Management Challenges in IAEA's Technical Cooperation Program*, GAO-09-275 (Washington, D.C.: Mar. 5, 2009).

[25] "Reinforcing the Global Nuclear Order for Peace and Prosperity: The Role of the IAEA to 2020 and Beyond," prepared by an independent commission at the request of the Director General of the IAEA (May 2008).

[26] As of the time we disseminated our survey, the Convention had not yet entered into force for two other countries, Libya and the United Arab Emirates, and we could not send our survey to a country for which it had not entered into force.

[27] We informed respondents that GAO is not authorized to withhold information from Congress, but that we received a written agreement from our congressional requester that he would not ask for individually identifiable survey information.

In: International Policies and Issues of Nuclear Safety... ISBN: 978-1-61761-596-2
Editor: Matthew D. Leeland

Chapter 3

SURVEY OF NATIONAL PROGRAMS FOR MANAGING HIGH-LEVEL RADIOACTIVE WASTE AND SPENT NUCLEAR FUEL

United States Nuclear Waste Technical Review Board

ACKNOWLEDGMENTS

The Board wishes to express its appreciation to the following individuals and institutions for the help and advice they provided in the preparation of the report.

In-Country Technical Reviewers

Georg Arens: Federal Ministry for the Environment, Nature Conservation, and Nuclear Safety (Germany)

Peter De Preter: National Agency for Radioactive Waste and Enriched Fissile Materials (Belgium)

Jo-Ann Facella: Nuclear Waste Management Organization (Canada)

Thomas Flüeler: Federal Institute of Technology (Switzerland)

Erik Frank: Federal Nuclear Safety Inspectorate (Switzerland)

Atika Khan: Nuclear Waste Management Organization (Canada)

Matti Kojo: University of Tampere (Finland)

Simon Löw: Federal Institute of Technology and National Council for Nuclear Waste (Switzerland)

John Mathieson: Nuclear Decommissioning Authority (United Kingdom) Claudio Pescatore: Nuclear Energy Agency, Organization for Economic Cooperation and Development

Esko Ruokola: Radiation and Nuclear Safety Authority (Finland)

Eva Simic: Swedish National Council for Nuclear Waste

Mitsuo Takeuchi: Nuclear Waste Management Organization (Japan)

Ju Wang: China National Nuclear Corporation

Jooho Whang: Kyung Hee University (Republic of Korea)
Staff of the National Agency for Radioactive Waste Management (France)

In addition, reviewers from Belgium (1), Finland (1), Germany (2), Spain (1), Sweden (1), and the United Kingdom (1) asked to remain anonymous.
Any errors that remain are the responsibility of the Board.

Credits for Cover Photographs (from back to front, left to right)

Swedish Nuclear Fuel and Waste Management Corporation: CLAB
U.S. Department of Energy: Aerial view of Yucca Mountain
French National Agency for Radioactive Waste Management: Underground Research Laboratory at Bure
Posiva Oy (Finland): Waste package
U.S. Department of Energy: Exploratory Studies Facility at Yucca Mountain
The Board appreciates the permission granted by the International Atomic Energy Agency to use portions of its **Radioactive Waste Management Glossary**, *2003 Edition, Publication 1155, (IAEA: Vienna, 2003).*

OVERVIEW

The creation of high-activity, long-lived radioactive waste is an inevitable consequence of generating electricity in nuclear power plants. It also is an inevitable consequence of engaging in a set of activities associated with national defense, ranging from propelling nuclear submarines to producing the fissionable materials needed to construct nuclear weapons. Early in the nuclear era, the very-longterm management and the ultimate disposition of those wastes was not a high priority. By the mid-1970s, however, most nuclear-capable nations had begun to focus more intently on developing plans to ensure over the very long term that the wastes would not endanger public health and safety or do serious damage to the environment.

Especially within the last decade, those efforts have benefitted from increasingly fruitful international cooperation and coordination. The International Atomic Energy Agency, an autonomous organization with a working relationship with the United Nations, carries out technical assistance programs and provides regulatory guidance to its members. It also supports the implementation of the *Joint Convention on the Safety of Spent Fuel Management and on the Safety of Radioactive Waste Management,* a treaty signed by 55 entities, including all but five of the 31 nations that now operate nuclear power plants. Under the auspices of its Radioactive Waste Management Committee, the Nuclear Energy Agency, part of the Organization for Economic Cooperation and Development, has established a number of subsidiary bodies that sponsor multinational exchanges among 28 industrial democracies.[1]

Today, there is strong international consensus that a deep geologic repository used to dispose of high-activity, long-lived radioactive waste "provides a unique level and duration of protection"[2] of public health and safety and the environment. Such a system "takes advantage of the capabilities of both the local geology and the engineered materials to fulfill specific safety functions in a complementary fashion providing multiple and diverse barrier roles." Further, the international waste management community broadly agrees that developing a deep geologic repository is "technically feasible." However, the route and pace in moving toward deep underground disposition of high-activity, long-lived radioactive waste vary considerably among countries with nuclear programs. Only one deep geologic repository is operating today: the Waste Isolation Pilot Plant in New Mexico. Transuranic waste from the U.S. nuclear weapons production program is the sole material that can be disposed of in that facility.

The purpose of this chapter is to provide Congress, the Secretary of Energy, and other interested parties with up-to-date information on the status of selected national programs to manage high-activity, long-lived radioactive waste.[3] The report is not intended to provide a comprehensive and exhaustive survey of waste management programs in the 31 countries that now operate nuclear power plants. Instead the report examines programs in 13 selected countries, which account for 83 percent of worldwide nuclear power generating capacity. These countries illustrate well the broad range of options and considerations that structure national programs. Importantly, all of these efforts are relatively transparent, thereby engendering some confidence that the information provided here is reliable. Other countries that might have been selected were ultimately omitted from this survey because their programs are in their infancy or because the status of their programs could not be independently documented. In the future, the Board may update this survey and include additional national programs.

For each of the 13 national programs, the Board gathered detailed information on 30 program attributes. Some of the attributes address the programs' legal and institutional arrangements; others describe technical approaches that the programs have taken. (A definition of these attributes can be found starting on page 13.) These data are presented in a series of detailed tables, which were reviewed for accuracy by at least one and in some cases as many as three in-country experts. The rest of this section highlights the following program attributes.[4]

- Context
- Organizational form of the implementer
- Independent technical/program oversight
- Current practices
- Geological investigations
- Status of the site-selection process
- Health and safety requirements for disposal
- Anticipated start of repository operations

Context

Commercial and defense spent nuclear fuel (SNF), high-level radioactive waste (HLW), long-lived heat- generating waste, wastes from research and isotope production reactors are all potential candidates for disposal in a deep geologic repository. But the magnitude of this task is best approximated by the capacity of a country's commercial nuclear power plants.

The 13 countries considered in the survey vary significantly in that respect (Table 1). Belgium, Finland, Spain, and Switzerland have few nuclear power plants; France, Japan, and the United States have a substantial number. Further, the countries' dependence on nuclear ranges from a small per- centage of national electricity production to a large majority of it.

In some nations, commitments have been made to construct a large number of new reactors, while, in others, the commitments have been more mod- est. In still others, legal or *de facto* moratoria on building nuclear power plants are in place. Finally, in many nations, a linkage exists, sometimes informal and implicit, sometimes formal and explicit, between finding a "solution" to the radioactive waste management problem and continued operation (or new construction) of nuclear power plants.

Organizational Form of the Implementer

A broad international consensus exists among countries actively considering the very-long-term management of radioactive waste that establishing health, safety, and environmental standards for disposal and deciding whether a deep geologic repository should be sited, constructed, or operated are intrinsic governmental functions to be carried out by an independent regulator.[5] There is considerably less agreement on what is the most appropriate organizational form for the implementing entity responsible for repository siting, construction, and operation. In the United States, for example, even as Congress gave implementing responsibility to the Department of Energy, it authorized the creation of a special commission to make recommendations about alternative means for financing and managing that responsibility.[6] No particular organizational form dominates national choices (Table 2). Although the language that individual countries use to describe the organizational form varies, four distinct types of organizations have been created: government agencies, private corporations, government-owned corporations, and public-private partnerships.

Table 1. Nuclear-Generated Electricity*

COUNTRY	OPERATING NUCLEAR POWER PLANTS	CURRENT GENERATING CAPACITY (GIGAWATTS)	PERCENTAGE OF TOTAL ELECTRICITY PRODUCTION
United States	104	101.1	19.7
Belgium	7	5.7	53.8
Canada	18	12.7	14.8
China	11	8.6	2.2
Finland	4	2.7	29.7
France	58	63.5	76.2
Germany	17	20.3	28.3
Japan	53	46.2	24.9
Republic of Korea	20	17.7	35.6
Spain	8	7.4	18.3
Sweden	10	9.1	42.0
Switzerland	5	3.2	39.2
United Kingdom	19	11.0	13.5

*As of May 31, 2009.

Table 2. Organizational form of the Implementer

COUNTRY	IMPLEMENTING ORGANIZATION	ORGANIZATIONAL FORM
United States	Department of Energy Office of Civilian Radioactive Waste Management	Government agency
Belgium	National Agency for Radioactive Waste and Enriched Fissile Materials	Government agency
Canada	Nuclear Waste Management Organization	Private corporation formed by the owners of nuclear fuel waste
China	China National Nuclear Corporation, provisionally	Government-owned corporation
Finland	Posiva Oy	Joint waste management company created by the owners of nuclear power plants
France	National Agency for Radioactive Waste Management	Government-owned Public Service Agency
Germany	Office for Radiation Protection within the Federal Ministry for the Environment, Nature Conservation, and Nuclear Safety	Government agency
Japan	Nuclear Waste Management Organization	Private nonprofit organization established by the owners of nuclear power plants
Republic of Korea	Korea Radioactive Waste Management Organization	Government agency
Spain	Spanish National Company for Radioactive Waste	Government-owned corporation
Sweden	Swedish Nuclear Fuel and Waste Management Company	Private corporation formed by the owners of nuclear power plants
Switzerland	National Cooperative for the Disposal of Radioactive Waste	Public/private consortium of radioactive waste producers, including the owners of nuclear power plants and the Federal State
United Kingdom	Nuclear Decommissioning Authority Radioactive Waste Management Division	Non-Departmental Public Body under the responsibility of the Department of Energy and Climate Change

Independent Technical/Program Oversight

In addition to the implementer and the regulator, in many countries, a third type of organization has been created: the independent technical/program oversight body (Table 3). These organizations can make findings and recommendations to the responsible governmental agencies and branches of government or to the implementer; but they have no authority or control over either the implementer or the regulator. Some of these bodies consciously were established to bolster the credibility of other organizations charged with programmatic responsibilities. Others were created to institutionalize a "second opinion" into what are often technically and politically controversial activities. Further, these oversight bodies differ in their charters. Some focus exclusively on technical matters while others have a broader mandate, which includes waste management's ethical, legal, social, and policy dimensions as well as its technical ones.

Table 3. Independent Technical/Program Oversight

COUNTRY	OVERSEER	ROLE
United States	Nuclear Waste Technical Review Board	Advises Congress and the Secretary of Energy
Belgium	None	
Canada	Independent Technical Review Group	Advises the Nuclear Waste Management Organization
China	No decision made.	
Finland	None	
France	National Review Board	Advises Government and Parliament
	Local Information and Oversight Committee	Advises National Agency for Radioactive Waste on the operation of the underground research laboratory at Bure
Germany	Nuclear Waste Management Commission	Advises the Ministry for the Environment, Nature Conservation, and Nuclear Safety
Japan	None	
Republic of Korea	None	
Spain	None	
Sweden	National Council for Nuclear Waste	Advises the Ministry of the Environment
Switzerland	Nuclear Safety Commission	Advises the Federal Council, the Department of the Environment, Transport, Energy, and Communication, and the Federal Nuclear Safety Inspectorate
United Kingdom	Committee on Radioactive Waste Management	Advises Government and the Devolved Administration Ministers

Table 4. Current Practices

COUNTRY	MATERIAL AUTHORIZED TO BE DISPOSED OF IN A DEEP GEOLOGIC REPOSITORY	INDEPENDENT CENTRALIZED STORAGE FACILITY ESTABLISHED
United States*	High-level radioactive waste, commercial spent nuclear fuel, naval reactor fuel, Department of Energy spent nuclear fuel	No
Belgium	High-level radioactive waste	No
Canada	Commercial spent nuclear fuel	No
China	High-level radioactive waste	No
Finland	Commercial spent nuclear fuel	No
France	High-level radioactive waste and long-lived intermediate-level waste	No
Germany	High-level radioactive waste, commercial spent nuclear fuel, and heat-generating intermediate-level waste	Yes, at Gorleben and Ahaus
Japan	High-level radioactive waste	No
Republic of Korea	Commercial spent nuclear fuel	Envisioned, but no schedule for beginning to site such a facility has been established
Spain	No decision made.	A siting process for a Centralized Temporary Storage facility has been initiated.
Sweden	Commercial spent nuclear fuel	Yes, at Oskarshamn (CLAB)
Switzerland	High-level radioactive waste is the only form currently considered waste.	Yes, at Würenlingen (ZWILAG)
United Kingdom	High-level radioactive waste, commercial spent nuclear fuel, intermediate-level waste, and low-level waste not suitable for near-surface disposal. Uranium and plutonium if these elements are declared to be waste.	No

*Does not include waste authorized to be disposed of in the Waste Isolation Pilot Plant.

Current Practices

Until the mid-1970s, the nuclear power community worldwide believed that the fuel assemblies from commercial nuclear power plants would be removed when spent, cooled on-site for a relatively short period of time, and then sent to chemical reprocessing plants. At the reprocessing plants, the plutonium, created by neutron absorption, and the remaining uranium would be separated out and recycled into either light-water or fast reactors. The HLW from reprocessing plants would be vitrified and ultimately disposed of in a deep geologic repository. Adopting this approach meant that the nuclear fuel cycle was "closed."

In part out of concerns about nuclear-weapon materials proliferation, some nations subsequently adopted a nuclear fuel cycle that was "once-through"—that is, the SNF removed from reactors would be cooled, stored either on-site or at an independent centralized facility, and then disposed of directly in a deep geologic repository.

There is considerable variety in materials authorized for disposal in different countries (Table 4). The variety is, to a large degree, a reflection of nuclear-fuel-cycle choices, some of which are still in flux. Countries also vary in whether they have established an independent centralized storage facility either for HLW or SNF or for both forms of waste.

Geological Investigations

An influential 1957 report by the U.S. National Academy of Sciences' National Research Council identified geological disposal as a technically defensible option for the very-long-term management of HLW. Further, the report pointed to salt as an acceptable host rock because the presence of salt implied the absence of water and because the plasticity of salt would seal fractures that otherwise could preferentially conduct water flow through the waste emplacement zones.[7] This chapter provided the technical rationale that led the United States to focus almost exclusively until the mid-1970s on identifying a site in salt for a deep geologic repository.

Finding a salt formation where a repository might be sited was not an option, however, for many other countries. Sweden, for example, began to explore the possibility of disposing of its waste in granite, which underlies most of that country's landscape. Out of that exploration eventually came the KBS-3 approach: SNF first is encapsulated in copper, and the copper canisters then are placed in granite basement rock at a depth of about 500 meters and surrounded by bentonite clay. The acceptance by the radioactive waste disposal community that a *combination* of geological and engineered barriers might provide suffi cient protection of public health and safety and the environment opened the door to investigate a large number of potential host rocks.[8] Indeed, even countries that are thinking about or actually developing a deep geologic repository in clay, and thus believe there is no need for a robust engineered barrier system, take a systems approach in developing their safety cases.

Numerous types of rock have been considered or have been investigated, and many countries have constructed an indigenous underground research laboratory to carry out in situ investigations of a formation's potential to isolate and contain radioactive waste (Table 5).

Table 5. Geological Investigations

COUNTRY	GEOLOGIC ENVIRONMENTS CONSIDERED OR INVESTIGATED	INDIGENOUS UNDERGROUND RESEARCH LABORATORY ESTABLISHED
United States	Salt, basalt, granite, tuff, clay, and shale	The Exploratory Studies Facility at Yucca Mountain served the function of an underground research laboratory (tuff).
Belgium	Clay and shale	Mol (clay)
Canada	Granite and sedimentary rock	Pinawa (granite)*
China	Granite	None
Finland	Granite, gneiss, grandiorite, and migmatite	Construction of ONKALO underground rock characterization facility in Eurajoki began in 2004 and is continuing (granite).
France	Argillite and granite	Bure (argillite)
Germany	Salt	Gorleben (salt)
Japan	Granite and sedimentary rock	Tono (granite) Mizunami (granite) Horonobe (sedimentary rock)
Republic of Korea	Granite	Korea Underground Research Tunnel (granite)**
Spain	Granite, clay, and salt	None
Sweden	Granite	Äspö (granite)
Switzerland	Clay and granite	Mont Terri (clay) and Grimsel (granite)
United Kingdom	No decision made.	None

*In the process of being decommissioned
**At shallow depth only

Status of the Site-selection Process

Experience over the years in many countries has made clear that potential deep geologic repository sites have to pass through both a technical filter and a political filter. Some countries identify potential sites based first on technical considerations and then determine whether political realities will permit the site's development as a repository. Other countries reverse the order, looking first for volunteer communities and then evaluating a site's technical merits. Still other countries have concluded that moving forward at this time is simply premature.

Table 6. Status of the Site-Selection Process

COUNTRY	STATUS
United States	Site at Yucca Mountain was selected.*
Belgium	Formal siting process not initiated.
Canada	Siting process initiated.
China	Preliminary investigations underway at Beishan in the Gobi Desert.
Finland	Site at Olkiluoto near the municipality of Eurajoki has been selected.
France	Site near the village of Bure has been selected.**
Germany	Siting process on hold.
Japan	Siting process initiated.
Republic of Korea	Formal siting process not initiated.
Spain	Formal siting process not initiated.
Sweden	Site in the municipality of Östhammar was selected.
Switzerland	Siting process initiated.
United Kingdom	Siting process initiated.

* The Administration recently indicated its intention to terminate funding for the Yucca Mountain repository program and to appoint a Blue Ribbon Commission to consider nuclear waste management alternatives. To date, licensing hearings before the Nuclear Regulatory Commission continue.

** A 250-square-kilometer area has been identified. The selection of a specific location within that area for development as a deep geologic repository is under way.

Experience over the years also has made clear that a siting process can get bogged down because either technical or political obstacles have arisen. In some countries, programs have had to be altered fundamentally to overcome either barrier or both barriers. These reorganizations have, at the very least, resulted in significant programmatic delays. In other countries, technical or political controversies, or the prospects of them, have lead policy-makers to defer for many decades the development of a deep geologic repository. Table 6 provides information on the status of the site-selection process in the 13 nations considered here.

Health and Safety Requirements for Disposal

Although a deep geologic repository can provide a unique level and duration of protection, questions remain in many countries about what the protective level should be, how standards should be formulated, how long the duration should last, what methodology should be used to judge compliance, and what the spatial domain should be where the regulation s enforced. Further, in some countries, the regulations are very prescriptive, and in others, they are very general, providing only broad guidelines for the implementer.

Table 7. Health and Safety Requirements for Disposal

COUNTRY	DOSE CONSTRAINT	RISK LIMIT	COMPLIANCE PERIOD
United States	0.15 mSv/year	Not specified	Less than 10,000 years
	1.0 mSv/year*	Not specified	Greater than 10,000 years but less than 1,000,000 years
Belgium	Expected to be 0.1 -0.3 mSv/year	Not specified	May be as much as 1,000,000 years
Canada	An upper limit of 1.0 mSv/year established; 0.3 mSv/year proposed.	Not specified	Not specified
China	No decision made.	No decision made.	No decision made.
Finland	Less than 0.1 mSv/year. Release limits for various radionuclides established.	Not specified	First several thousand years
	Impacts should be comparable to those arising from natural radioactive materials but should remain insignificantly low.	Not specified	Beyond first several thousand years.
France	0.25 mSv/year for normal scenarios.	Not specified	10,000 years
Germany	Not specified	Less than 10^{-4}/lifetime for probable scenarios; Less than 10^{-3}/lifetime for less probable scenarios	1,000,000 years
Japan	No decision made.	No decision made.	No decision made.
Republic of Korea	No decision made.	No decision made.	No decision made.
Spain	No decision made.	No decision made.	No decision made.
Sweden	Not specified	Less than 10^{-5}/year	100,000 years
Switzerland	Complete containment	Not specified	1,000 years
	0.1 mSv/year for probable scenarios	Not specified	As much as 1,000,000 years
	Not specified	Less than 10^{-6}/year for less probable scenarios	As much as 1,000,000 years
United Kingdom	No decision made.	Guidance calls for less than 10^{-6}/year.	No decision made.

*Applicable only to a repository constructed at Yucca Mountain.

Three aspects of the regulations seem to be of particular importance. In the terminology used by the radioactive waste management community, the minimal acceptable protective level is measured by either a dose constraint or a risk limit. The dose constraint is the effective dose or the equivalent dose to individuals that may not be exceeded. The dose constraint is usually measured in millisieverts per year.[9] No consensus obtains on the definition of a risk limit. Typically, however, the term is taken to mean the probability of a person living in the vicinity of a repository suffering genetic or serious health effects, including cancer, during the course of his or her life-time as a result of radioactive material released from the repository.[10] The risk limit is always measured in terms of the probability per year, for example, one in a million or 10–6/year. Finally, the duration over which the regulation applies is measured by a compliance period. These regulatory choices represent social judgments informed by technical analyses.[11]

Some of the 13 nations discussed in this chapter have not yet established radiological health and safety requirements for the disposition of radioactive waste. Among those that have, there are some important similarities and differences (Table 7). If one looks only at the first 10,000 years after repository closure, all the countries regard as acceptable a dose constraint that falls within a range of 0.1–0.3 mSv/year. However, some of those countries require that the dose constraint also be satisfied for compliance periods that extend as far out as 1,000,000 years. Risk limits span a similarly large range—10-3 to 10-6/year—depending on the compliance period and the likelihood that a particular scenario evolves.

Table 8. Anticipated Start of Repository Operations

COUNTRY	DATE
United States	No decision made.*
Belgium	Anticipated in roughly the 2040 time-frame.
Canada	No decision made.
China	Anticipated in roughly the 2050 time-frame.
Finland	2020
France	2025
Germany	No decision made.
Japan	No decision made.
Republic of Korea	No decision made.
Spain	No decision made.
Sweden	2023
Switzerland	No sooner than 2040
United Kingdom	No decision made.

* The Administration recently indicated its intention to terminate funding for the Yucca Mountain repository program and to appoint a Blue Ribbon Commission to explore nuclear waste management alternatives. To date, licensing hearings before the Nuclear Regulatory Commission continue.

Anticipated Start of Repository Operations

Three of the four countries that have selected a site for a deep geologic repository—Finland, France, and Sweden—have announced when they anticipate the start of operations: they all expect to begin emplacing radioactive waste within roughly the next 10 to 15 years (Table 8). Three other countries have projected a time 30 to 40 years from now, although only one of them, Switzerland, has initiated a formal siting process. The remaining seven nations have made no decision about when repository operations will begin, either because the timing depends on finding a volunteer community and reaching an agreement with it or because a formal siting process is on hold or has not been initiated.

A Concluding Comment

When the radioactive waste management community in the United States and abroad began work on developing a deep geologic repository, the task was perceived to be a simple one. A technically suitable site would first be identified. Then scientific and engineering talents would be mustered to complete what was viewed as a relatively straightforward construction project. Since then, it has become clear that performing convincing technical analysis in the face of considerable temporal and spatial uncertainties is more complex and challenging than earlier anticipated. Creating a supportive institutional environment—founding credible implementing and regulatory agencies, creating trusting relationships with local communities, and putting into place legitimate decision-making processes—has proven to be challenging as well. That many national programs have had to be reconstituted in fundamental ways is testimony to the diffi culties encountered over the years.

With only the three countries identified above close to implementing a technically and politically accepted effort to develop a deep geologic repository, it is diffi cult to infer what, if anything, is a "magical recipe" for success. This question, however, will be explored in greater depth in a subsequent Board report. At this point, we simply note that these 13 countries are strongly committed to managing radioactive waste for the very long term in ways that do not impose burdens on future generations. The precise path taken by each will strongly depend on its technical and political cultures. In the end, it may very well be that the many paths all lead to the same outcome: successful disposal of long-lived, highactivity radioactive waste in a deep geologic repository.

DETAILED TABLES

The Board identified 30 key attributes associated with national radioactive waste management programs. Half of them relate to the institutional arrangements that have been established in each country. The other attributes relate to the technical approaches that have been adopted. Detailed tables containing information about the attributes, which appear as column headings, were then constructed using offi cial documents released by each nation. Most helpful were a series of reports submitted to the International Atomic Energy Agency as part of the Third Meeting held under the provisions of the *Joint Convention on the Safety of Spent Fuel Management and the Safety of Radioactive Waste Management.*

Drafts of the detailed tables were sent out to in-country experts for peer review. At least one expert from each country, and as many as three, reviewed the tables and provided comments. (Reviewers that wished to be recognized are acknowledged in the beginning of this chapter.) The comments were incorporated, and the tables were revised. Typically, the comments filled in "blanks" on the draft tables, updated the information contained in the drafts, and provided more information than could be found in the offi cial documents. Every effort was made to harmonize the various table entries so that their meaning would be consistent across countries. To achieve that end, definitions for the 30 attributes were developed. These definitions are provided below. [12]

To facilitate the viewing of the detailed tables, they are laid out so that the information for three countries is grouped together. The attributes related to institutional arrangements for each group are presented first followed by the attributes related to technical approaches. With the exception of the United States, the nations are grouped in alphabetical order. Each group is color-coded so that the reader can quickly locate any country of interest.

Key Attributes—Institutional Arrangements

1. *Legislation specific to radioactive waste management:* Laws passed that establish the rules under which radioactive waste will be managed.
2. *Implementing organization:* The entity charged under the law with the responsibility for siting, constructing, and operating facilities for managing radioactive waste.
3. *Independent regulator:* The entity charged under the law with the responsibility for establishing health, safety, and environmental standards for managing radioactive waste and for approving/disapproving or recommending for approval/disapproval the licensing of facilities for managing radioactive waste.
4. *Independent technical/program oversight:* Entities that are independent of the implementer and the regulator that provide advice on technical and other issues associated with the management of radioactive waste. The entities can give their advice to the Government, the legislature, or the implementer. They can be appointed either by the Government or the implementer.
5. *Dedicated funding source for repository development:* Money, segregated from general government revenues, that finances the siting, construction, and operation of a deep geologic repository and other facilities. The source of the money may be payments by waste generators directly or by the users of nuclear-generated electricity.
6. *Regulations and decrees applicable to licensing a deep geologic repository—site-selection:* Rules and standards created by government agencies and Ministries that structure the processes used to choose a candidate or final location for a deep geologic repository.
7. *Regulations and decrees applicable to licensing a deep geologic repository—environmental impact assessment:* Rules and standards created by government agencies and Ministries that structure the processes and the required analysis for evaluating the environmental effects of developing a deep geologic repository.

8. *Regulations and decrees applicable to licensing a deep geologic repository—health and safety protection:* Rules and standards created by government agencies and Ministries that structure the processes and the required analysis for evaluating whether a proposed deep geologic repository is likely to comply with applicable requirements for protecting public health and safety.
9. *Formal legislative/executive approvals required for developing a deep geologic repository —selection of a waste management option:* Decisions about whether to develop a deep geologic repository or to adopt some other option, such as separation and transmutation or storage, for the very-long-term management of radioactive waste. The decision is made using political, as opposed to administrative, processes. It may occur before or after a regulatory decision or the submission of regulatory advice to the legislature or Government.
10. *Formal legislative/executive approvals required for developing a deep geologic repository —site-selection:* The decision to choose a candidate or final location for a deep geologic repository. The decision is made using political, as opposed to administrative, processes. It may occur before or after a regulatory decision or the submission of regulatory advice to the legislature or Government.
11. *Formal legislative/executive approvals required for developing a deep geologic repository —facility construction and operation:* The decision to permit the construction and operation of a deep geologic repository. The decision is made using political, as opposed to administrative, processes. It may occur before or after a regulatory decision or the submission of regulatory advice to the legislature or Government.
12. *Interactions with local jurisdictions—local veto:* Legally prescribed rules under which either a locality must give its approval before an action is taken (usually the selection of a site for a deep geologic repository) or the locality can reject a decision after it has been made.
13. *Interactions with local jurisdictions—limitations on local veto:* Legally prescribed rules under which any veto power held by local jurisdictions can be overridden or otherwise modified.
14. *Interactions with local jurisdictions—benefits to be provided to local community for accepting a facility:* Benefits include, among other things, dedicated tax and other payments, increased governmental services, and infrastructure development. Benefits may be legally prescribed or established through negotiations.
15. *Explicit adoption of a staged decision-making process:* Almost by necessity, the development of a deep geologic repository must take place in stages. However, some national programs are designed to require at every step intensive deliberation, recursive safety case evaluations, and explicit consideration of the option of not proceeding.

Key Attributes—Technical Approaches

1. *Operating nuclear power plants/generating capacity:* The number and gross generating capacity (in gigawatts electric) of plants operating as of May 31, 2009.

The number of plants under construction is defined as the projects that have broken ground as of May 31, 2009. The generating capacity of those plants is the nominal capacity reported to authorities.

2. *Reprocessing included in fuel cycle:* Whether SNF has ever been reprocessed, either in the country or in a facility located outside of the country.
3. *Transportation system in place to move SNF/HLW to a deep geologic repository:* Transportation options available for those countries where a site has been selected or where particular sites are being actively considered.
4. *Independent centralized interim storage facility established:* Facilities that fall into this category store SNF or HLW from more than one generator. Such facilities are distinguished from storage installations at either operating or shut-down nuclear power or reprocessing plants.
5. *Geologic environments considered or investigated for a repository:* Host rocks that appear to be potentially suitable for a repository. The hosts rocks may have been considered or investigated in bench or desk studies, by surface investigation, or by at-depth exploration.
6. *Indigenous underground research laboratories:* Laboratories that have been created (either operational or under construction) where experiments can be conducted to evaluate the long-term suitability of a particular host rock to isolate and contain radioactive waste. Experiments conducted in another country's underground research laboratory are not included.
7. *Status of site-selection process:* The stage of the national decision for selecting a site for a deep geologic repository.
8. *Long-term health and safety requirements:* Specific regulations and standards establishing dose constraints, risk limits, and compliance periods that must be satisfied before a deep geologic repository can be licensed.
9. *Requirements for retrievability:* Specific laws or regulations establishing the time period within which waste must be able to be retrieved from a deep repository. Also specific laws and regulations establishing how the entire disposal process can be reversed.
10. *Requirements for defense-in-depth:* Specific laws or regulations establishing the degree to which various barriers must be able to isolate and contain radioactive waste independently of other barriers.
11. *Methodology for demonstrating compliance with postclosure standards:* Approaches the implementer must use to conduct its performance assessment or to advance its safety case for licensing a deep geologic repository.
12. *Engineered barrier system—design:* How the man-made part of the deep geologic repository system is to be constructed.
13. *Engineered barrier system—importance to safety case:* In comparison to the natural system (host rock, near-field environment, hydrogeology, and other factors), the role of the engineered barrier system in isolating and containing radioactive waste.
14. *Waste forms authorized to be disposed of in a deep geologic repository:* The type of material that would be required to be disposed of in a deep geologic repository.
15. *Anticipated start of repository operations:* Year in which either the implementer or an appropriate governmental authority has stated publicly that a deep geologic repository will be available to begin to accept waste for disposal.

UNITED STATES, BELGIUM, CANADA

Institutional Arrangements

	Legislation Specific to Radioactive Waste Management	Implementing Organization	Independent Regulator	Independent Technical/ Program Oversight	Dedicated Funding Source for Repository Development
United States*	Nuclear Waste Policy Act (1982) Nuclear Waste Policy Amendments Act (1987) Energy Policy Act (1992)	Department of Energy Office of Civilian Radioactive Waste Management (Government agency)	Environmental Protection Agency (Sets environmental standards) Nuclear Regulatory Commission (Implements standards and licenses facilities)	Nuclear Waste Technical Review Board (Advises Congress and the Secretary of Energy)	Nuclear Waste Fund Generators of nuclear electricity pay a $0.001 per kilowatt-hour surcharge into the Fund.
Belgium	Law of 8 August 1980, modified by Law of 11 January 1991 and Law of 12 December 1997 [Implementation] Law of 29 March 1958, modified by Law of 15 April 1994 [Regulation]	National Agency for Radioactive Waste and Enriched Fissile Materials (Government agency)	Federal Agency for Nuclear Control	None	Long-Term Fund Costs of developing a repository will be fully paid by waste generators.
Canada	Nuclear Fuel Waste Act (2002) Nuclear Safety and Control Act (2000)	Nuclear Waste Management Organization, subject to Government approval of key policies and decisions (Private corporation formed by the owners of nuclear fuel waste)	Canadian Nuclear Safety Commission	Independent Technical Review Group (Advises the Nuclear Waste Management Organization)	Nuclear Fuel Waste Act Trust Fund Owners of nuclear power plants pay money into the Fund, subject to the formula approved by Government.

(Continued)

	Regulations and Decrees Applicable to Licensing a Deep Geologic Repository			Formal Legislative/Executive Approvals Required for Developing a Deep Geologic Repository		
	Site-Selection	**Environmental Impact Assessment**	**Health and Safety Protection**	**Selection of Waste Management Option**	**Site-Selection**	**Facility Construction and Operation**
United States*	Nuclear Regulatory Commission 10CFR60 Department of Energy 10CFR960 and 10CFR963 (Radioactive waste-specific)	Council of Environmental Quality 40CFR1500 (Generic)	Nuclear Regulatory Commission 10CFR60 and 10CFR63 Environmental Protection Agency 40CFR191 and 40CFR197 (Radioactive waste-specific)	Department of Energy Generic Environmental Impact Statement (1980) Nuclear Waste Policy Act (1982)	Congressional approval of the Yucca Mountain site recommendation by President (2002)	After approval by the Nuclear Regulatory Commission of licenses to construct and possess/receive waste, no further action is required.
Belgium	Decision by the Council of Ministers limiting siting activities for a low- and intermediate-level waste repository to nuclear or non-nuclear volunteering sites No specific law for HLW/SNF repository (Radioactive waste-specific)	Strategic Environmental Assessment (Law 13 of February 2006) (Radioactive waste-specific)	Federal Agency for Nuclear Control Royal Decree of 20 July 2001 GRR-2001 (Radioactive waste-specific)	Strategic Environmental Assessment (Law 13 of February 2006) triggered the development of a Waste Plan by National Agency for Radioactive Waste and Enriched Fissile Materials, but no decision has been made.	No decision made.	Construction and operating licenses granted by Government through a Royal Decree on the advice of Federal Agency for Nuclear Control.
Canada	Canadian Nuclear Safety Commission Regulations under the Nuclear Safety and Control Act Clauses 3 and 4 (Nuclear facility-specific) Geological Considerations in Siting a Repository for High-Level Radioactive Waste R-72 (Radioactive-waste specific)	Regulations under the Canadian Environmental Assessment Act (Generic)	Canadian Nuclear Safety Commission Regulatory Guide G-320 (Radioactive waste-specific)	Government approved a recommendation by the Nuclear Waste Management Organization to implement a geologic disposal strategy of Adaptive Phased Management. An optional strategy of shallow underground storage also was adopted. (2007)	Once the Federal Minister of the Environment approves the environmental assessment, the Canadian Nuclear Safety Commission has the full authority to decide on the Site Preparation License.	After the approval by the Canadian Nuclear Safety Commission of licenses to construct and possess/receive waste, no further action is required.

(Continued)

	Interactions with Local Jurisdictions			Explicit Adoption of a Staged Decision-Making Process
	Local Veto	Limitations on Local Veto	Benefits to be Provided to Local Community for Accepting a Facility	
United States*	Yes, by governor	State veto can be exercised only when President recommends site for Congress's approval. Veto can be overridden by majority vote in both Houses of Congress	A schedule for providing benefits to the State of Nevada and to any state or tribe hosting a centralized interim storage facility or a repository was included in the Nuclear Waste Policy Amendments Act. Nevada has never requested benefits.	No
Belgium	No decision made for HLW/SNF repository. No formal legal right of veto for low- and intermediate-level waste disposal site. But without the agreement of the local partnership, the disposal facility will not be built. Volunteer nuclear community (Dessel) agreed to host a low- and intermediate-level waste disposal site.	No decision made for HLW/SNF repository. "Gentleman's agreement" that a low- and intermediate-level waste disposal facility will not be sited without community consent. As the project develops, implementer can veto decisions by the local community that adversely affect safety.	No decision made for HLW/SNF repository. None provided to date for low- and intermediate-level waste disposal facility. Discussions are on-going about providing benefits in the future.	No decision made. Expectation is that the process will be flexible and iterative.
Canada	Under the Adaptive Phased Management strategy, only communities willing to host a geologic repository will be considered.	No decision made.	No decision made.	Yes

* Does not include institutional arrangements for the Waste Isolation Pilot Plant.

Technical Approaches

	Operating Nuclear Power Plants/ Generating Capacity	Reprocessing Included in Fuel Cycle	Transportation System in Place to Move SNF/HLW to a Deep Geologic Repository	Independent Centralized Interim-Storage Facility Established	Geologic Environments Considered or Investigated for a Repository
United States*	104 nuclear power plants (101.1 GWe) One nuclear power plant is under construction. (1.2 GWe)	The U.S. reprocessed SNF as part of its weapons plutonium production program. Small amounts of commercial SNF were reprocessed at West Valley, N.Y. Two other commercial reprocessing plants were constructed but never operated. The U.S. does not currently reprocess commercial SNF.	Depends on where the repository is developed. No rail transportation system is available for the Yucca Mountain site.	No	Salt, basalt, tuff, granite, clay, and shale
Belgium	Seven nuclear power plants (5.7 GWe) Additional nuclear power plants and operation beyond 40 years for existing nuclear power plants were prohibited in 2003.	Commercial SNF was reprocessed at La Hague. Moratorium on new reprocessing contracts was instituted in 1993 and confirmed in 1998 by the Council of Ministers. A small amount of commercial SNF was reprocessed by the pilot facility Eurochemic in Dessel.	No decision made.	No	Clay and shale
Canada	18 nuclear power plants (12.7 GWe)	No	No decision made.	No	Granite or sedimentary rock

(Continued)

	Indigenous Underground Research Laboratories	Status of Site-Selection Process	Long-Term Health and Safety Requirements	Requirements for Retrievability	Requirements for Defense-in-Depth
United States*	The Exploratory Studies Facility at the Yucca Mountain site served the function of an underground research laboratory. (Tuff)	The Yucca Mountain site has been characterized and was approved by Congress in 2002. The Administration has recently indicated its intention to terminate funding for the Yucca Mountain repository program and to appoint a Blue Ribbon Commission to explore nuclear waste management alternatives. Licensing hearings before the Nuclear Regulatory Commission continue.	For Yucca Mountain: 0.15 mSv/year for 10,000 years; 1 mSv/year thereafter up until 1,000,000 years. For any other site: 0.15 mSv/year for 10,000 years.	Within 50 years from the start of waste emplacement	Multiple barriers (both natural and engineered) required. No requirement for defense-in-depth or redundancy.
Belgium	HADES Project initiated in 1974 in Mol. (Clay)	No active siting process for a HLW/SNF repository being carried out.	No decision made. Dose constraint expected to be in the range 0.1–0.3 mSv/year. Compliance period may be as long as 10^6 years	No decision made.	No decision made. Safety philosophy includes defense-in-depth and some degree of redundancy in barrier function.
Canada	Pinawa Laboratory in Manitoba. (Granite) (In the process of decommissioning)	The Nuclear Waste Management Organization has proposed a process for selecting a site. No schedule has been set to complete the site-selection process.	Canadian Nuclear Safety Commission has specified a public dose limit of 1 mSv/year. Implementer is required to provide rationale for dose constraint. An example of a proposed dose constraint is 0.3 mSv/year.	The Adaptive Phased Management plan includes "potential for retrievability of the used fuel for an extended period, until such time as a future society makes a determination on the final closure and the appropriate form and duration of postclosure monitoring." This requirement has not yet been incorporated into regulations.	The Adaptive Phased Management plan recognizes the value of multiple barriers and redundant systems. This recognition has not yet been translated into regulations.

(Continued)

	Methodology for Demonstrating Compliance with Postclosure Standards	Engineered Barrier System		Waste Forms Authorized to be Disposed of in a Deep Geologic Repository	Anticipated Start of Repository Operations
		Design	Importance to Safety Case		
United States*	Mean value of Monte Carlo realizations generated by a probabilistic Total System Performance Assessment	Double-shelled waste package composed of Alloy 22 (outer) and carbon steel (inner); titanium drip shield	Very important	Vitrified commercial and defense HLW, commercial SNF, Navy SNF, and DOE SNF	No decision made.
Belgium	No decision made. A host-rock specific (Boom clay) performance assessment (SAFIR-2) has been carried out (2001): • Evaluation of normal scenario and altered scenarios • Some probabilistic elements • 0.3 mSv/year dose constraint • Calculations carried out to at least 10^6 years	Current reference design developed after SAFIR-2 considers stainless steel canisters holding HLW and a carbon steel overpack surrounded by thick concrete. The so-called Supercontainer is placed in concrete-lined drifts and backfilled with cementitious materials.	Relatively unimportant, although preliminary safety assessments do not support conclusively the isolation and containment capacity of Boom clay. (SAFIR-2)	Only HLW because SNF is not considered "waste." However, National Agency for Radioactive Waste and Enriched Fissile Materials must study geological disposal for both HLW and SNF.	No decision made. Likely to be in the 2040 time-frame.
Canada	According to the Canadian Nuclear Safety Commission Regulatory Guide G-320, an applicant to site, construct, or operate a geologic repository can choose among the following methodologies in developing its safety case: • Scoping assessments • Bounding assessments • Realistic best estimates of performance • Conservative calculations • Deterministic or probabilistic calculations	No decision made.	No decision made.	Commercial SNF	No decision made.

* Does not include institutional arrangements for the Waste Isolation Pilot Plant.

China, Finland, France

Institutional Arrangements

	Legislation Specific to Radioactive Waste Management	Implementing Organization	Independent Regulator	Independent Technical/ Program Oversight	Dedicated Funding Source for Repository Development
China	Law of the People's Republic of China on Prevention and Control of Radioactive Pollution (2003)	China National Nuclear Corporation (Government-owned corporation)	National Nuclear Safety Administration within the Ministry of Environmental Protection	No decision made.	No decision made.
Finland	Nuclear Energy Act (1987) Nuclear Energy Decree (1988) Nuclear Energy Act Amendments (1994, 2003, 2008)	Posiva Oy (Joint waste management company created in 1995 by two utilities, Fortum Power and Heat Oy and Teollisuuden Voima Oyj.)	Radiation and Nuclear Safety Authority (Advises Government on the safety of proposed facilities)	None	Nuclear Waste Management Fund Generators estimate cost of radioactive waste disposal and nuclear power plant decommissioning. They pay annually the difference between Fund target and amount existing in the Fund. Payment can be in securities. Excess payments can be recovered.
France	Research on Radioactive Waste Management Act (1991) Planning Act Concerning the Sustainable Management of Radioactive Materials and Waste (2006) Transparency and Security in the Nuclear Field (2006)	National Agency for Radioactive Waste Management reporting to the Ministries of Environment, Industry, and Research (Government-owned Public Service Agency)	Nuclear Safety Authority	National Review Board (Advises Government and Parliament) Local Information and Oversight Committee (Must be consulted on issues relating to the operation of an underground research laboratory.)	Yes The National Agency for Radioactive Waste Management must estimate costs of designing, constructing, operating, and closing a repository. The waste generators must contribute to a Fund, which is supervised by an independent commission established under the 2006 Planning Act.

(Continued)

	Regulations and Decrees Applicable to Licensing a Deep Geologic Repository			Formal Legislative/Executive Approvals Required for Developing a Deep Geologic Repository		
	Site-Selection	Environmental Impact Assessment	Health and Safety Protection	Selection of Waste Management Option	Site-Selection	Facility Construction and Operation
China	National Nuclear Safety Administration Guidelines on Siting of a Radioactive Waste Geological Repository HAD-406/06 (Radioactive waste-specific)	Decree on Environmental Protection (Generic)	National Nuclear Safety Administration Regulations on the Safety Control for Civilian Nuclear Installations HAF001 Regulations on Radioactive Waste Safety HAF401 (Radioactive waste-specific)	Law of the People's Republic of China on Prevention and Control of Radioactive Pollution (2003)	No decision made.	No decision made.
Finland	Government Decree 736-2008 on the Safety of the Disposal of Nuclear Waste Radiation and Nuclear Safety Authority Long-term Safety of Disposal of SNF YVL 8.4 (Radioactive waste-specific)	Decree on Environmental Impact Assessment Procedures (Generic)	Government Decree 736-2008 on the Safety of the Disposal of Nuclear Waste Radiation and Nuclear Safety Authority Long-term Safety of disposal of SNF YVL 8.4 Operation of the final disposal facility for SNF YVL 8.5 (Radioactive waste-specific)	Nuclear Energy Act (1987, 1994, 2008)	Decision-in-Principle by Government (2000) Confirmation of Decision-in-Principle by Parliament (2001, 2002)	Construction and operating licenses are granted by the Government on the advice of Radiation and Nuclear Safety Authority.
France	Nuclear Safety Authority Safety Guide for Final Disposal of Radioactive Waste In Deep Geologic Formations (Radioactive waste-specific)	Code of Environment Articles L121 and R121-R125 (Generic)	Nuclear Safety Authority Safety Guide for Final Disposal of Radioactive Waste In Deep Geologic Formations (Radioactive waste-specific)	Research on Radioactive Waste Management Act (1991)	Approval of a site near the village of Bure in the Meuse/Haute-Marne region in the Planning Act Concerning the Sustainable Management of Radioactive Materials and Waste (2006).	After a license application is submitted, Parliament will prescribe reversibility conditions that have to be met. Afterward, the Council of State can grant the license.

(Continued)

	Interactions with Local Jurisdictions			Explicit Adoption of a Staged Decision-Making Process
	Local Veto	Limitations on Local Veto	Benefits to be Provided to Local Community for Accepting a Facility	
China	No decision made.	No decision made.	No decision made.	No decision made.
Finland	Yes, by Municipal Council. The Eurajoki nuclear community approved a positive statement, thereby not vetoing the siting of a HLW/SNF repository.	Veto shall be exercised before the Government makes a Decision-in-Principle on the repository. Veto cannot be overridden.	A benefits package was negotiated between Eurajoki Township and Posiva Oy and Teollisuuden Voima Oyj in 1999. The scale of benefits is minimal, including a loan to construct a new home for the elderly. The former home was renovated and rented to Posiva Oy.	No
France	None, but the local governments in the Meuse/Haute-Marne region volunteered for an underground site-characterization program.	Not applicable	The 2006 Planning Act defines a series of measures to support local development, including a dedicated tax on Basic Nuclear Installations.	Yes

Technical Approaches

	Operating Nuclear Power Plants/ Generating Capacity	Reprocessing Included in Fuel Cycle	Transportation System in Place to Move SNF/HLW to a Deep Geologic Repository	Independent Centralized Interim-Storage Facility Established	Geologic Environments Considered or Investigated for a Repository
China	11 nuclear power plants (8.6 GWe) 16 nuclear power plants are under construction. (16.0 GWe)	Yes A small reprocessing plant located in northwest China is expected to begin operation in 2010.	No decision made.	No	Granite, shale, tuff, mudstone, and diorite
Finland	Four nuclear power plants (2.7 GWe) One nuclear power plant is under construction. (1.6 GWe)	No	No final decision has been made about transportation mode. Readily available options include sea, truck, and rail.	No	Granite, gneiss, granodiorite, and migmatite
France	58 nuclear power plants (63.5 GWe) One nuclear power plant is under construction. (1.6 GWe)	Yes	Meuse/Haute-Marne region is accessible only by truck, but studies are under way to explore alternatives.	No	Argillite and granite

(Continued)

	Indigenous Underground Research Laboratories	Status of Site-Selection Process	Long-Term Health and Safety Requirements	Requirements for Retrievability	Requirements for Defense-in-Depth
China	None	Preliminary investigations are under way at the Beishan site (granite) in the Gobi Desert in Gansu Province in Northwest China. Site-Selection is not anticipated before 2020.	No decision made.	No decision made.	No decision made.
Finland	Construction of ONKALO underground rock characterization facility in Eurajoki began in 2004 and is continuing. Experimental work is being conducted during construction. (Migmatite)	Olkilouto, a site at Eurajoki in migmatite, has been approved by Government (2000) and by Parliament (2001). It is being characterized at depth.	For the first several thousand years, dose limit is less than 0.1 mSv/year for normal events. Release limits for selected radionuclides established. Beyond the first several thousand years, impacts can be comparable to those arising from natural radioactive substances but should remain insignificantly low.	The regulatory requirement for retrievablility was eliminated in 2008. However, Posiva is still required under the 2000 Decision-in-Principle to present a plan and cost estimate for retrieving the waste when it submits an application for a construction license.	The barriers should complement each other so that a deficiency in one will not jeopardize long-term safety.
France	Construction of the Meuse/Haute-Marne facility near the village of Bure began in 1999. (Argillite)	The National Agency for Radioactive Waste Management is deciding, in consultation with local communities, where in the transposition zone, a $250km^2$ area north of Bure, the repository should be sited.	Dose limit is 0.25 mSv/year for normal scenarios. Compliance period is 10^4 years.	The repository must be designed so that it is "reversible" for at least 100 years. Reversibility is a management concept that requires technical retrievability.	Required Safety Guide for Final Disposal of Radioactive Waste In Deep Geologic Formations Chapters 5.1 and 6.1

(Continued)

	Methodology for Demonstrating Compliance with Postclosure Standards	Engineered Barrier System		Waste Forms Authorized to be Disposed of in a Deep Geologic Repository	Anticipated Start of Repository Operations
		Design	Importance to Safety Case		
China	No decision made.	No decision made.	No decision made.	HLW	Around 2050
Finland	Compliance is to be demonstrated by means of a deterministic, conservative safety case that addresses both the expected evolutions and unlikely disruptive events affecting long-term safety. The safety case consists of a numerical analysis based on experimental studies and will be complemented by qualitative expert judgment whenever quantitative analyses are not feasible or are too uncertain.	Double-shelled waste package composed of copper (outer) and cast-iron (inner); the annulus between the canister and the rock wall will be filled with highly compacted bentonite.	Very important	Commercial SNF	2020
France	Compliance is shown through the deterministic evaluation of several normal and altered scenarios. In addition, deterministic sensitivity calculations are used to evaluate the impact of uncertainty.	Vitrified waste placed within stainless steel packages.	Minimal	HLW and long-lived intermediate-level waste	2025

GERMANY, JAPAN, REPUBLIC OF KOREA

Institutional Arrangements

	Legislation Specific to Radioactive Waste Management	Implementing Organization	Independent Regulator	Independent Technical/ Program Oversight	Dedicated Funding Source for Repository Development
Germany	Atomic Energy Act (1959) Nuclear Licensing Procedure Ordinance (1977) Federal Mining Act (1980) Waste Disposal Advance Payments Ordinance (1982) Precautionary Radiation Protection Act (1986) Radiation Protection Ordinance (2001)	Federal Office for Radiation Protection within the Federal Ministry for the Environment, Nature Conservation, and Nuclear Safety may make use of "third parties," such as the private German Service Company for the Construction and Operation of Waste Repositories (Government Agency)	Supervisory authority and author of rules and guidelines: Federal Ministry for the Environment, Nature Conservation, and Nuclear Safety Licensing authorities: States (Länder) Ministries under the Atomic Energy Act, State authorities for water use and protection, and competent State mining authorities under the Federal Mining Act	Nuclear Waste Management Commission (Advises the Ministry for the Environment, Nature Conservation, and Nuclear Safety)	None, but research and development costs for a repository are recovered from the waste generators.
Japan	Final Disposal of Specific Radioactive Wastes Act (2000, 2007) Reactor Regulation Law (1957, amended in 2007 to establish a safety regulation system for the disposal of HLW and transuranic waste)	Nuclear Waste Management Organization (Private nonprofit organization established by the owners of nuclear power plants)	Nuclear and Industrial Safety Agency, a unit within the Ministry of Economy, Trade, and Industry	None, although the Nuclear Safety Commission advises the Nuclear and Industrial Safety Agency.	Yes. With respect to HLW, each year, the Ministry of Economy, Trade, and Industry notifies nuclear power plant owners how much they need to deposit in the High-Level Waste Fund. Yes. With respect to Transuranic Waste, the Ministry of Economy, Trade, and Industry notifies the owners of reprocessing plants and mixed-oxide fuel fabrication plants how much they need to deposit in the Transuranic Waste fund.
Republic of Korea	Atomic Energy Act (1988) Enforcement Decree of the Atomic Energy Act (1988) Radioactive Waste Management Law (2008)	Korea Radioactive Waste Management Organization (Government Agency)	Ministry of Education, Science, and Technology (Advised by Korea Institute of Nuclear Safety and Nuclear Safety Commission)	None	"Polluter pays" principle adopted in Atomic Energy Act Radioactive Waste Management Fund No decision made on implementation except that the generator pays only after waste is accepted for centralized storage or disposal.

(Continued)

	Regulations and Decrees Applicable to Licensing a Deep Geologic Repository			Formal Legislative/Executive Approvals Required for Developing a Deep Geologic Repository		
	Site-Selection	**Environmental Impact Assessment**	**Health and Safety Protection**	**Selection of Waste Management Option**	**Site-Selection**	**Facility Construction and Operation**
Germany	No decision made.	Environmental Impact Assessment Act (Generic)	Federal Ministry for the Environment, Nature Conservation, and Nuclear Safety Safety Requirements Governing the Final Disposal of Heat-Generating Radioactive Waste (Radioactive waste-specific)	Atomic Energy Act (1959)	No decision made.	Plan Approval by the State where the site is located, subject to supervision by the Federal Government.
Japan	None The Nuclear Safety Commission has published guidelines (not regulations) that the Nuclear Waste Management Organization must follow.	No decision made.	Federal Ministry for the Environment, Nature Conservation, and Nuclear Safety Rules for Category 1 Waste Disposal for Nuclear Fuel Material (Radioactive waste-specific)	Final Disposal of Specific Radioactive Wastes Act (2000, revised in 2007)	With the consent of the Cabinet, the Minister of Economy, Trade, and Industry must give approval.	With consent of the Cabinet, the Minister of Economy, Trade, and Industry must give approval.
Republic of Korea	Ministry of Education, Science, and Technology Siting Criteria for Spent Fuel Interim Storage Facilities Notice Number 2008-58 No decision made for HLW/SNF repository. (Radioactive waste-specific)	Ministry of Education, Science, and Technology Regulation on the Environmental Radiation Survey and Impact Analysis in the Vicinity of Nuclear Facilities Notice Number 2008-28 (Radioactive waste-specific)	Ministry of Education, Science, and Technology Criteria for Radioactive Dose Notice Number 1998-12 and 2005-17 promulgated by the Ministry of Education, Science, and Technology's predecessor body, the Ministry of Science and Technology (Radioactive waste-specific)	Atomic Energy Act (1988)	No decision made.	No decision made.

(Continued)

	Interactions with Local Jurisdictions			Explicit Adoption of a Staged Decision-Making Process
	Local Veto	Limitations on Local Veto	Benefits to be Provided to Local Community for Accepting a Facility	
Germany	No decision made.	No decision made.	No decision made.	"Stepwise optimization" is mandated under the Safety Requirements Governing the Final Disposal of Heat-Generating Radioactive Waste.
Japan	The Mayor of host community and the Prefectual Governor must agree to participate in the siting process.	None	If a local community agrees to be included in a literature survey of potential sites, it and its neighboring communities will receive up to $18 million. If the community subsequently allows surface-based site investigations, it and its neighboring communities will receive up to $65 million.	Yes
Republic of Korea	No decision made for HLW/SNF repository. A volunteer nuclear community (Kyongju/Gyeonju) agreed in a referendum to host a low- and intermediate-level waste disposal site.	No decision made on HLW/SNF repository. None for a low- and intermediate-level waste disposal site.	No decision made for HLW/SNF repository. $300 million provided local community for low- and intermediate-level waste disposal site; in addition, the community will receive $10 million per year after the facility begins operation. Other benefits are possible.	No decision made for HLW/SNF disposal. A staged decision-making process for low- and intermediate-level waste disposal is in place.

Technical Approaches

	Operating Nuclear Power Plants/ Generating Capacity	Reprocessing Included in Fuel Cycle	Transportation System in Place to Move SNF/HLW to a Deep Geologic Repository	Independent Centralized Interim-Storage Facility Established	Geologic Environments Considered or Investigated for a Repository
Germany	17 nuclear power plants (20.3 GWe)	Before 1994, commercial SNF had to be reprocessed. Between January 1, 1994, and June 30, 2005, nuclear power plant owners had the option of reprocessing their commercial SNF. Under amendments to the Atomic Energy Act in 2002, transport of commercial SNF to reprocessing plants after July 1, 2005, was prohibited. Most of the reprocessing of German commercial SNF was done at La Hague, although smaller amounts were reprocessed at Sellafield, EUROCHEMIC, and Karlsruhe.	No decision made.	Facilities at Gorleben and Ahaus store small amounts of commercial SNF. HLW is stored at Gorleben. Under the amendments to the Atomic Energy Act in 2002, commercial SNF has to be stored at nuclear power plants.	Salt
Japan	53 nuclear power plants (46.2 GWe) Three nuclear power plants are under construction. (3.7 GWe)	Yes. Commercial SNF from Japan has been reprocessed in France and the United Kingdom. In addition, reprocessing takes place in a small facility at Tokai and, pending the results of pre-service testing, will take place at a large facility at Rokkasho Village.	No decision made.	No	Granite and sedimentary rock
Republic of Korea	20 nuclear power plants (17.7 GWe) Six nuclear power plants are under construction. (6.7 GWe)	All SNF used in light-water reactors is of U.S. origin. The U.S. has refused to permit the reprocessing of that fuel.	No decision made.	Envisioned, but no schedule for beginning to site such a facility has been established.	Granite

(Continued)

	Indigenous Underground Research Laboratories	Status of Site-Selection Process	Long-Term Health and Safety Requirements	Requirements for Retrievability	Requirements for Defense-in-Depth
Germany	Underground exploration of the Gorleben site was launched in 1986 but was suspended in 2000. (Salt)	The Site-Selection process, which focused on the Gorleben site, was suspended in 2000 for no longer than 10 years. The suspension was agreed to by the Government and the nuclear power plant owners. The Federal Ministry for the Environment, Nature Conservation, and Nuclear Safety published a concept for the site-selection process in 2006.	The additional risk of serious health damage must be less than10^{4}/lifetime for probable scenarios.* The additional risk of serious health damage must be less than 10^{-3}/lifetime for less probable scenarios.* Compliance period is 10^{6} years. * lifetime means 100 years for humans	Retrievability of HLW and SNF need not be provided in the disposal concept. However, shielding of the ionizing radiation has to be guaranteed so that the waste will be manageable for possible retrieval for a period of 500 years after repository closure, taking into account probable developments.	The safety after repository closure has to be secured through a robust and graded barrier system. The functions of this system should be fulfilled passively and without any maintenance. In addition, the functionality should be maintained in a sufficient way even for the case that individual barriers will not provide their full effectiveness.
Japan	Tono (granite) Two laboratories are under construction: Mizunami (granite) and Horonobe (sedimentary rock).	The Nuclear Waste Management Organization has adopted a transparent and voluntary approach for finding candidate sites. One town (Toyo-cho) initially agreed to participate but then withdrew. The national government may take a more proactive role in identifying sites in the future.	No decision made.	No decision made.	No decision made.
Republic of Korea	Korea Underground Research Tunnel at shallow depth (Granite)	No schedule for beginning to site a HLW/SNF repository has been established. A volunteer nuclear community (Kyongju/Gyeonju) has agreed to host a low- and intermediate-level waste disposal site.	No decision made for HlW/ SNF disposal. For low- and intermediate-level waste disposal, 0.1 mSv/year for normal events; risk limit of 10^{-6}/year for probabilistic disruptive events; 1 mSv/year for human intrusion.	No decision made, but, in the conceptual-level Reference Design, waste packages had to be retrievable for an indeterminate period.	Safety philosophy includes defense-in-depth and some degree of redundancy in multi-barrier function.

(Continued)

	Methodology for Demonstrating Compliance with Postclosure Standards	Engineered Barrier System		Waste Forms Authorized to be Disposed of in a Deep Geologic Repository	Anticipated Start of Repository Operations
		Design	Importance to Safety Case		
Germany	Deterministic calculations have to be carried out on the basis of modeling as realistically as possible, using, for example, median values as input parameters.	No decision made.	No decision made.	HLW, commercial SNF, and heat-generating intermediate-level waste	No decision made.
Japan	No decision made. A generic performance assessment (H12) has been carried out (1999): • No specific host rock modeled. • A range of scenarios evaluated deterministically. • Sensitivity calculations addressed uncertainty. • Calculations carried out to at least 10^8 years. • Peak dose was less than 0.1mSv/year.	No decision made. In the H12 performance assessment, vitrified waste in stainless steel containers is disposed of in a carbon steel overpack that is surrounded by bentonite.	No decision made. In the H12 performance assessment, all packages were assumed to fail at 10^3 years.	HLW	No decision made.
Republic of Korea	No decision made. A quantitative approach for up to 10,000 years is envisioned, with qualitative assessments carried out for subsequent periods.	No decision made. A conceptual-level Reference Design was developed in the late 1990's. In the Reference Design, options considered for the canisters were high-nickel alloys, stainless steel, and copper. The canisters would be surrounded by bentonite.	In the conceptual-level Reference Design, the engineered barrier system was very important.	Commercial SNF	No decision made.

SPAIN, SWEDEN, SWITZERLAND

Institutional Arrangements

	Legislation Specific to Radioactive Waste Management	Implementing Organization	Independent Regulator	Independent Technical/ Program Oversight	Dedicated Funding Source for Repository Development
Spain	Nuclear Energy Act, Law 25/1964, modified by Law 24/2005 Act Creating the Nuclear Safety Council, Law 15/1980, modified by Law 33/2007	Spanish National Company for Radioactive Waste (Government-owned Corporation)	Nuclear Safety Council The Ministry of Industry, Tourism, and Trade legally makes the final decision but cannot overturn the Nuclear Safety Council's report if it is negative or conditional.	None	Nuclear Decommissioning Fund paid into by the waste producers. (Covers both decommissioning of nuclear power plants and radioactive waste management)
Sweden	Act on Nuclear Activities (1984) Radiation Protection Act (1988) Environmental Code (1998)	Swedish Nuclear Fuel and Waste Management Company (Private corporation formed by the owners of nuclear power plants)	Radiation Safety Authority, within the Ministry of the Environment, was established in 2008 by merging the Radiation Protection Institute and the Nuclear Power Inspectorate.	National Council for Nuclear Waste (Advises the Ministry of the Environment)	Nuclear Waste Fund Owners of nuclear power plants pay a fee based on the estimated costs of disposing of SNF. The fee varies from year to year and varies as well from plant to plant. Owners provide a guarantee to cover the difference between money paid into the Fund and the total estimated cost of disposal.
Switzerland	Atomic Energy Act (1959, modified to the Nuclear Energy Act in 2003) Environmental Protection Act (1983) Radiological Protection Act (1991) Radiological Protection Ordinance (1994) Ordinance on the Collection of Radioactive Waste (2002) Nuclear Energy Ordinance (2004) Nuclear Safety Inspectorate Act (2007) Ordinance on Decommissioning and Waste Management Fund (2008)	National Cooperative for the Disposal of Radioactive Waste (The National Cooperative for the Disposal of Radioactive Waste is a public/private consortium of radioactive waste producers, including all the owners of nuclear power plants and the Federal State.)	Department of Environment, Transport, Energy, and Communications, advised by the Federal Nuclear Safety Inspectorate and the Federal Nuclear Safety Commission	Nuclear Safety Commission (Advises Federal Council, Federal Department of Environment, Transport, Energy, and Communications, and the Federal Nuclear Safety Inspectorate)	Radioactive Waste Disposal Fund for Nuclear Installations Consumers of nuclear-generated electricity pay into the Fund a surcharge for each kilowatt-hour produced. Current expenses incurred by the National Cooperative for the Disposal of Radioactive Waste are paid for annually by the nuclear power plant owners.

(Continued)

	Regulations and Decrees Applicable to Licensing a Deep Geologic Repository			Formal Legislative/Executive Approvals Required for Developing a Deep Geologic Repository		
	Site-Selection	Environmental Impact Assessment	Health and Safety Protection	Selection of Waste Management Option	Site-Selection	Facility Construction and Operation
Spain	Royal Decree 775/2006 established process for siting the Centralized Temporary Storage facility. No decision has been made for a HLW/SNF repository.	Royal Decree 1/2008	Royal Decree 1836/1999, modified by Royal Decree 35/2008 Nuclear Safety Council Regulation on Nuclear and Radioactive Facilities	No decision made.	No decision made.	No decision made.
Sweden	None	The Environmental Impact Assessment process is consistent with Directives from the European Union.	Regulations promulgated by the Radiation Protection Institute and the Nuclear Power Inspectorate have been adopted by the Radiation Safety Authority. SSIFS 1998 :1 SSIFS 2005:5 SKIFS 1998:1 SKIFS 2002 :1 SKIFS 2004: 1	In 1983, the Government decided that the KBS-3 approach formed an acceptable basis for the Swedish Nuclear Fuel and Waste Management Company to use in its plans for developing a repository.	In 2001, the Government approved a proposal by the Swedish Nuclear Fuel and Waste Management Company to investigate sites in Östhammar and Oskarshamn to determine whether they were suitable for developing a repository.	The Radiation Safety Authority will review a license application, scheduled to be submitted in 2010, and advise the Government on its acceptability. Concurrently, an Environmental Court will rule on the application. Taking the Radiation Safety Authority's advice and the Environmental Court's ruling into account, the Government will decide whether to approve the license application. There also may be a non-binding vote of Parliament.
Switzerland	Federal Council in 2008 approved the concept of the Sectoral Plan (Sachplan).	Federal Office for the Environment reviews Environmental Impact Assessments.	Federal Nuclear Safety Inspectorate Protection Objectives for the Disposal of Radioactive Waste Guideline G03 (2009) replaces HSK-R-21 (1993).	Nuclear Energy Acts of 1959 and 2003	General License granted by Federal Council and approved by Parliament. General License may be challenged in a national Facultative Referendum.	Construction and operating licenses granted by Federal Department of Environment, Transport, Energy, and Communications.

(Continued)

	Interactions with Local Jurisdictions			Explicit Adoption of a Staged Decision-Making Process
	Local Veto	Limitations on Local Veto	Benefits to be Provided to Local Community for Accepting a Facility	
Spain	No decision made for a HLW/SNF repository. The process established for siting the Centralized Temporary Storage facility requires voluntary participation by local communities.	No decision made for a HLW/SNF repository.	No decision made for a HLW/SNF repository.	No decision made for a HLW/SNF repository.
Sweden	Local community can veto the choice of a site.	National override of veto only if there is no alternative location for the repository in a community more willing to accept it.	Oskarshamn, the community not selected, will receive, in an added-value package, $180 million for participating in the siting process. Östhammar, the community selected, will receive $60 million.	A license for a geological repository for SNF will be granted in two steps. The first step involves the full licensing of a facility. After several years of trial operation, regular operation will begin.
Switzerland	None, although informal participation and formal consultations are required at all stages of the Sectoral Plan.	Not applicable	No decision made. To be decided in Stage 3 of the Sectoral Plan.	The Sectoral Plan contains three stages.

Technical Approaches

	Operating Nuclear Power Plants/ Generating Capacity	Reprocessing Included in Fuel Cycle	Transportation System in Place to Move SNF/HLW to a Deep Geologic Repository	Independent Centralized Interim-Storage Facility Established	Geologic Environments Considered or Investigated for a Repository
Spain	Eight nuclear power plants (7.4 GWe)	Some commercial SNF from the Vandellós I plant was reprocessed at La Hague. Additional commercial SNF from the Santa María de Garoña plant was reprocessed at Sellafield. Current national policy does not contemplate any additional reprocessing.	No decision made.	A process was initiated in 2006 to site a Centralized Temporary Storage facility. No site has been chosen.	Granite, clay, and salt No geological environments are currently under consideration.
Sweden	10 nuclear power plants (9.1 GWe)	No Although small amounts of commercial SNF have been reprocessed in France and the United Kingdom, no vitrified waste was returned to Sweden.	Waste can be moved to the Östhammar site by a specially designed ship, the Sigyn.	Yes, the CLAB facility in Oskarshamn was commissioned in 1985.	Granite
Switzerland	Five nuclear power plants (3.2 GWe)	Commercial SNF has been reprocessed in France and the United Kingdom. Some HLW has been returned to Switzerland. Some extracted plutonium and some extracted uranium has been used to make MOX assemblies. Discharged SNF not covered by reprocessing contracts in place in 2002 may not be reprocessed until at least 2016.	No decision made.	A Central Storage Facility (ZWILAG) in Würenlingen, near the Beznau Nuclear Power Plant, holds HLW and SNF.	Clay and granite

(Continued)

	Indigenous Underground Research Laboratories	Status of Site-Selection Process	Long-Term Health and Safety Requirements	Requirements for Retrievability	Requirements for Defense-in-Depth
Spain	None	No decision made.	For low- and intermediate-level waste disposal, dose constraint is 0.1 mSv/year for high-probability scenarios; risk limit of 10^{-6}/year is for lower probability scenarios. No decision made for HLW/SNF repository.	No decision made.	No decision made.
Sweden	Construction of the Äspö laboratory in Oskarshamn began in 1990 and was completed in 1995. (Granite)	A site in the municipality of Östhammar was selected in 2009.	Risk limit of 10^{-6}/year for 10^5 years. The risk analysis should be carried out for no longer than 10^6 years.	None	The barrier system shall comprise several barriers so that, as far as possible, the necessary safety is maintained in spite of a failure in one barrier.
Switzerland	Mont Terri near Saint Ursanne (clay) and Grimsel in the Berne Canton (granite)	Sectoral Plan is at Stage 1, in which potential regions are examined based on their geological characteristics. The National Cooperative for the Disposal of Radioactive Waste has recommended three regions for further exploration. All are in Opalinus clay and are in northern Switzerland. The proposals are under review by Department of Environment, Transport, Energy, and Communications, advised by the Federal Nuclear Safety Inspectorate, the Federal Nuclear Safety Commission, and the Commission for Nuclear Waste Disposal.	Dose constraint relevant to high-probability scenarios 0.1 mSv/year; risk target of 10^{-6}/year for lower-probability scenarios. Complete containment is required for 1,000 years. The protection of man and the environment is "permanent." However, as a practical matter, safety assessments will be carried out for 10^7 years.	The retrievability of HLW has to be considered when designing the repository. The technical feasibility of retrieving the waste has to be demonstrated in experiments on a 1:1 scale before the repository starts operation.	The regulations require that the long-term safety of a repository shall be ensured by a system of multiple passive safety barriers. Although some "redundancy to ensure insensitivity to uncertainties" is required, the regulations do not specify any particular level.

(Continued)

	Methodology for Demonstrating Compliance with Postclosure Standards	Engineered Barrier System		Waste Forms Authorized to be Disposed of in a Deep Geologic Repository	Anticipated Start of Repository Operations
		Design	Importance to Safety Case		
Spain	No decision made.	No decision made.	No decision made.	No decision made.	No decision made.
Sweden	The regulations do not prescribe a specific methodology for demonstrating compliance. Both deterministic and probabilistic approaches can be used. Three types of scenarios are to be evaluated: • Main scenario–based on the probable evolution of the external conditions using realistic or pessimistic assumptions. • Less probable scenarios–prepared for the evaluation of uncertainties. Include variations on the main scenario with alternative sequences of event. • Residual scenarios–include sequences of events and conditions that illustrate the significance of individual barriers and barrier functions.	Commercial SNF is placed in a copper canister that has a cast-iron insert for support. The canister is surrounded by bentonite clay.	Very important	Commercial SNF	2023
Switzerland	No decision made. In a study, Project Entsorgungsnachweis, evaluating the feasibility of the Opalinus clay disposal concept, a deterministic performance assessment of a "Reference Scenario" was carried out. In addition, five other ("what-if") scenarios, often containing subscenarios, were evaluated to explore the effects of uncertainty. Complementing the deterministic calculations were some probabilistic analyses.	The disposal concept evaluated in Project Entsorgungsnachweis envisions canisters for HLW and commercial SNF made out of cast iron. HLW is contained in a stainless steel flask inside the canister. The canisters are surrounded by bentonite clay.	Relatively unimportant compared to the Opalinus clay.	HLW	Not earlier than 2040

United Kingdom

Institutional Arrangements

	Legislation Specific to Radioactive Waste Management	Implementing Organization	Independent Regulator	Independent Technical/ Program Oversight	Dedicated Funding Source for Repository Development
United Kingdom	Nuclear Installations Act (1965) Radioactive Substances Act (1993)	Nuclear Decommissioning Authority Radioactive Waste Management Directorate (Non-Departmental Public Body under the responsibility of the Department of Energy and Climate Change; for some aspects of its functions, it is also responsible to Scottish Ministers.)	Environment Agency (for England and Wales) Scottish Environment Agency Northern Ireland Department of Environment Health and Safety Executive Department for Transport	Committee on Radioactive Waste Management (Advises Government and ministers of the devolved administrations of Scotland, Wales, and Northern Ireland)	None Government will pay the costs of managing legacy waste. The policy of the Government is that owners and operators of new nuclear power plants set aside funds to cover their full share of waste management and disposal costs.

(Continued)

	Regulations and Decrees Applicable to Licensing a Deep Geologic Repository			Formal Legislative/Executive Approvals Required for Developing a Deep Geologic Repository		
	Site-Selection	Environmental Impact Assessment	Health and Safety Protection	Selection of Waste Management Option	Site-Selection	Facility Construction and Operation
United Kingdom	None	European Union Environmental Impact Assessment Act transposed in the United Kingdom by the Planning (Environmental Impact Assessment) Regulations	Environment Agency Geological Disposal Facilities on Land for Solid Radioactive Wastes: Guidance of Requirements for Authorisation (2009)	Department of Environment, Food, and Rural Affairs Response to the Report and Recommendations of the Committee on Radioactive Waste Management (2006) Department of Environment, Food, and Rural Affairs White Paper on Managing Radioactive Waste Safely (2008)	Government must ultimately approve the site. However, under the voluntarism and partnership approach, at each stage of site-selection the process, the decision-making body for the local community will decide whether to proceed with the next step.	After authorization by the Environment Agency, planning permission/ development consent, and licensing by the Health and Safety Executive, no further action is required.

	Interactions with Local Jurisdictions			Explicit Adoption of a Staged Decision-Making Process
	Local Veto	Limitations on Local Veto	Benefits to be Provided to Local Community for Accepting a Facility	
United Kingdom	The Government has committed to siting a repository only in a "willing" community.	The right of the community to withdraw from the process can be exercised until the point that underground construction and operations begin.	No decision made. The Government believes that any benefits package should be developed among the communities, Government, and the Nuclear Decommissioning Authority, taking account of local needs, affordability, and the value of money.	The Managing Radioactive Waste Safely program contains six stages.

Technical Approaches

	Operating Nuclear Power Plants/ Generating Capacity	Reprocessing Included in Fuel Cycle	Transportation System in Place to Move SNF/HLW to a Deep Geological Repository	Independent Centralized Interim-Storage Facility Established	Geologic Environments Considered or Investigated for a Repository
United Kingdom	19 nuclear power plants (11.0 GWe)	Yes, at Sellafield. The Government's position is that the decision to reprocess in the future should be left to the commercial judgment of the owners of the SNF. SNF is not classified as waste.	No decision made.	No	No decision made.

	Indigenous Underground Research Laboratories	Status of Site-Selection Process	Long-Term Health and Safety Requirements	Requirements for Retrievability	Requirements for Defense-in-Depth
United Kingdom	None	The Government has solicited expressions of interest from local governments and communities. So far, there have been three expressions of interest, from the Copeland Borough Council, from the Allendale Borough Council, and from the Cumbria Borough Council. All of these localities are close to the Sellafield site.	According to regulatory guidances, radiological risk from a disposal facility to a person representative of those at greatest risk should be consistent with a risk limit level of 10^{-6}/year.	No decision made, but planning and guidance require that the option of retrievability not be foreclosed.	A requirement of regulatory guidance.

(Continued)

	Methodology for Demonstrating Compliance with Postclosure Standards	Engineered Barrier System		Waste Forms Authorized to be Disposed of in a Deep Geologic Repository	Anticipated Start of Repository Operations
		Design	Importance to Safety Case		
United Kingdom	No decision made.	No decision made.	No decision made.	HLW, commercial SNF, intermediate-level waste, and low-level waste not suitable for near-surface disposal. Uranium and plutonium if these elements are declared to be waste.	No decision made.

GLOSSARY*

analysis, deterministic A simulation of the behavior of a system utilizing a single-valued set of parameters, events, and features. *See* also **analysis, probabilistic**.

analysis, probabilistic A simulation of the behavior of a system defined by parameters, events, and features whose values are represented by a statistical distribution. The analysis gives a corresponding distribution of results.

analysis, risk An analysis of possible events and their probabilities of occurrence together with their potential consequences.

argillite A compact rock derived from either claystone, siltstone, or shale, that is more indurated than its constituent source rock but less laminated and fissile than shale and lacking the cleavage of slate.

assessment, environmental impact An evaluation of radiological and nonradiological impacts of a proposed activity where the performance measure is overall environmental impact, including radiological and other global measures of impact on safety and environment.

assessment, performance An assessment of the performance of a system or subsystem and its implications for protection and safety at a planned or an authorized facility. This differs from safety assessment in that it can be applied to parts of a facility and does not necessarily require assessment of radiological impacts.

assessment, safety An analysis for evaluating the performance of an overall system and its impact where the performance measure is radiological impact or some other global measure of impact on safety.

backfill The material used to refill excavated parts of a repository (drifts, disposal rooms, or boreholes) during and after waste emplacement.

barrier A physical or chemical feature that prevents or delays the movement of radionuclides or other material between components in a system — for example, a waste repository. In general, a barrier can be an engineered barrier that is constructed or a natural geological, geochemical, or hydrogeological barrier.

---

* Most of these definitions have been taken from International Atomic Energy Agency, *Radioactive Waste Management Glossary, 2003 Edition*, Publication 1155, (IAEA: Vienna, 2003). The definitions of some terms have been altered to make them more applicable to this chapter, and other terms have been added. The IAEA is not responsible for those changes. Definitions of geologic terms are derived from the *American Geological Institute Glossary of Geology,* Third and Fourth Editions (AGI: Alexandria, VA, .1987 and 1997).

barriers, multiple Two or more natural or engineered barriers used to isolate radioactive waste in, and prevent radionuclide migration from, a repository. *See* also barrier.

basalt A dark-colored mafic igneous rock, commonly extrusive as lava flows or cones but also intrusive as dikes or sills.

bentonite A soft light-colored clay formed by chemical alteration of volcanic ash. Bentonite has been proposed for backfill and buffer material in many repositories.

characterization, site Detailed surface and subsurface investigations and activities at candidate disposal sites for obtaining information to determine the suitability of the site for a repository and to evaluate the long-term performance of a repository at the site.

clay A sediment composed of rock or mineral fragments smaller than 4 microns. Clays typically have relatively low permeability and relatively high capacity for sorption of positively charged chemicals.

closure Administrative and technical actions directed at a repository at the end of its operating lifetime — for example, covering the disposed of waste (for a near-surface repository) or backfilling and/or sealing (for a geological repository and the pas-sages leading to it) — and termination and completion of activities in any associated structures.

compliance period The length of time over which a repository is expected to satisfy either the dose constraint or the risk limit.

containment Methods or physical structures designed to prevent the dispersion of radio-active substances.

crystalline rock *See* **rock, crystalline**.

decommissioning Administrative and technical actions taken to allow the removal of some or all of the regulatory controls from a facility. This does not apply to a repository or to certain nuclear facilities used for mining and milling of radioactive materials, for which the term closure is used.

defense-in-depth The application of more than one protective measure for a given safety objective so that the objective is achieved even if one of the protective measures fails.

direct disposal Disposal of spent nuclear fuel as waste.

disposal Emplacement of waste in an appropriate facility without the intention of retrieval.

disposal facility Synonymous with "repository."

dose constraint The value of the effective dose or the equivalent dose to individuals from releases from a repository that may not be exceeded.

drift A horizontal or nearly horizontal mined opening.

engineered barrier system The designed, or engineered, components of a repository, including waste packages and other engineered barriers. *See* **also barrier**.

environmental impact statement A set of documents recording the results of an evaluation of the physical, ecological, cultural, and socioeconomic effects of a proposed facility (e.g., a repository), of a new technology, or of a new program.

fuel cycle All operations associated with the production of nuclear energy, including mining and milling; processing and enrichment of uranium or thorium; manufacture of nuclear fuel; operation of nuclear reactors; reprocessing of nuclear fuel; related research and development activities; and all related radioactive waste management activities including decommissioning.

fuel cycle, once-through Refers to the fuel cycle option where spent fuel is disposed of directly after use and is not reprocessed. *See* also direct disposal.

fuel, spent nuclear (SNF) Nuclear fuel removed from a reactor following irradiation that is not intended for further use in its present form because of depletion of fissile material, buildup of poison, or radiation or other damage.

geologic barrier *See* **barrier**.

geologic disposal *See* **repository, geologic**.

geologic repository *See* **repository, geologic**.

glass (waste matrix material) An amorphous material with a molecular distribution similar to that of a liquid but with a viscosity so great that its physical properties are those of a solid. Glasses used in the solidification of liquid high-level waste are generally based on a silicon-oxygen network. Additional network formers, such as aluminum, or modifiers, such as boron, lead to aluminosilicate or borosilicate glass.

granite Broadly applied, any holocrystalline quartz-bearing plutonic rock. The main components of granite are feldspar, quartz, and, as a minor essential mineral, mica. Granite formations are being considered as possible hosts for geological repositories.

groundwater Water that is held in rocks and soil beneath the surface of the earth.

heat-generating waste *See* **waste, heat generating**.

high-level waste (HLW) *See* **waste, high-level**.

host medium/rock *See* **rock, host**.

intermediate-level waste *See* waste, low- and intermediate-level.

implementing organization The entity charged under law (and its contractors) that undertakes the siting, design, construction, commissioning, and operation of a nuclear facility.

***in situ* testing** Tests to determine the characteristics of the natural system that are conducted within a geological environment that is essentially equivalent to the environment of an actual repository.

license An authorization issued by the regulatory body granting permission to perform specified activities related to a facility or an activity. The holder of a current license is termed a "licensee."

lithostatic pressure Pressure due to the weight of overlying rock and/or soil and water.

long-lived waste *See* waste, **long-lived.**

long-term In radioactive waste disposal, refers to periods of time that exceed the time during which active institutional control can be expected to last.

low- and intermediate-level waste *See* waste, **low- and intermediate-level**.

model A conceptual, analytical, or numerical representation of a system and the ways in which phenomena occur within that system, used to simulate or assess the behavior of the system for a defined purpose.

multiple barriers *See* **barriers, multiple**.

nuclear fuel cycle *See* fuel cycle.

nuclear waste *See* **waste, radioactive**.

once-through fuel cycle *See* **fuel cycle, once through**.

overpack A secondary (or additional) outer container for one or more waste packages, used for handling, transport, storage, or disposal.

package, spent fuel A vessel containing conditioned spent fuel in a form suitable for transport, storage, and disposal.

package, waste The waste form and any container(s) and internal barriers (e.g., absorbing materials and liners), prepared in accordance with the requirements for handling, transport, storage, and disposal.

postclosure The period of time following the closure of a repository and the decommissioning of related surface facilities. *See* also **closure**.

probabilistic analysis *See* **analysis, probabilistic**.

regulator An authority or a system of authorities designated by the government of a nation as having legal authority for conducting the regulatory process, including issuing authorizations, and thereby for regulating the siting, design, construction, commissioning, operation, closure, decommissioning, and, if required, subsequent institutional control of nuclear facilities or specific aspects thereof.

repository, deep geologic A facility for disposal of radioactive waste located under-ground (usually several hundred meters or more below the surface) in a geological formation intended to provide long-term isolation of radionuclides from the biosphere.

reprocessing A process or operation the purpose of which is to extract radioactive iso-topes from spent fuel for further use or to separate out various waste streams.

risk A multiattribute measure expressing hazard, danger, or chance of harmful or injurious consequences associated with actual or potential exposures. It reflects the probability that specific deleterious consequences may arise and the magnitude and character of such consequences.

risk limit The probability of a person living in the vicinity of a repository suffering genetic or serious health damage, including cancer, during the course of his or her life-time as a result of radioactive material released from the isolating rock zone.

rock A solid aggregate composed of naturally occurring substances including either one or more minerals, glasses, or organic matter.

rock, crystalline A generic term for igneous rocks and metamorphic rocks as opposed to sedimentary rocks. *See* also **granite**.

rock, host A geological formation in which a repository is located.

rock, igneous Rock or mineral that solidified from molten or partly molten material. This includes plutonic rock such as granite and volcanic rocks such as basalt.

rock, sedimentary A type of rock resulting from the consolidation of loose material that has accumulated in layers. The layers may be built up mechanically or by chemical precipitation.

safety assessment *See* **assessment, safety.**

safety case An integrated collection of arguments and evidence for demonstrating the safety of a facility. This will normally include a safety assessment but could also typically include independent lines of evidence and reasoning on the robustness and reliability of the safety assessment and the assumptions made therein.

salt In geology, generally used to refer to naturally occurring halite (sodium chloride).

scenario A postulated or assumed set of conditions or events. Scenarios are commonly used in performance assessments to represent possible future conditions or events to be modeled, such as the possible future evolution of a repository and its surroundings.

sedimentary rock *See* **rock, sedimentary**.

shale A consolidated clay rock that possesses closely spaced, well-defined laminae.

site The area containing, or under investigation of its suitability for, a nuclear facility (e.g., a repository). It is defined by a boundary and is under effective control of an operating organization.

site characterization *See* **characterization, site**.

site confirmation The final stage of the siting process for a repository. Site confirmation is based on detailed investigations of the preferred site that provide site-specific information needed for safety assessment.

site selection *See* **siting**.

siting The process of selecting a suitable disposal site. The process comprises the following stages: concept and planning; area survey; site characterization; site confirmation.

spent nuclear fuel (SNF) *See* **fuel, spent nuclear**.

spent nuclear fuel management All activities that relate to the handling or storage of spent nuclear fuel.

spent fuel package *See* **package, spent fuel.**

storage The holding of spent nuclear fuel or of radioactive waste in a facility that provides for its containment, with the intention of retrieval.

storage, interim *See* **storage**.

transuranic waste *See* **waste, transuranic**.

tuff A rock composed of compacted volcanic ash.

underground research laboratory A facility where *in situ* testing can take place.

vitrification The process of incorporating materials into a glass or glass-like form. Vitrification is commonly applied to the solidification of liquid high-level radioactive waste from the reprocessing of spent nuclear fuel. *See* **also glass**.

vitrified waste *See* **waste glass**.

waste Material in gaseous, liquid, or solid form for which no further use is foreseen.

waste, heat generating Radioactive waste that is suffi ciently radioactive that the decay heat significantly increases its temperature and the temperature of its surroundings. In practice, heat-generating waste is normally high-level waste, although some types of intermediate-level waste may qualify as heat-generating waste.

waste, high-level (HLW) The radioactive liquid containing most of the fission products and actinides present in spent fuel — which forms the residue from the first solvent extraction cycle in reprocessing — and some of the associated waste streams; this material following solidification; spent fuel (if it is declared a waste); or any other waste with similar radiological characteristics. Typical characteristics of HLW are thermal powers that are above about 2 kW/m3 and long-lived radionuclide concentrations exceeding the limitations for short-lived waste.

waste, intermediate-level *See* **waste, low- and intermediate-level.**

waste, long-lived Radioactive waste that contains significant levels of radionuclides with half-lives above 30 years.

waste, low- and intermediate-level Radioactive waste with radiological characteristics between those of waste exempted from regulation and high-level waste and spent nuclear fuel. They may be long-lived waste or short-lived waste. Many countries sub-divide this class in other ways — for example, into low-level waste and intermediatelevel waste or medium-level waste, often on the basis of waste acceptance requirements for near-surface repositories.

waste, radioactive Waste that contains or is contaminated with radionuclides at concen-trations or activities greater than clearance levels as established by the regulatory body. t should be recognized that this definition is purely for regulatory purposes and that material with activity concentrations equal to or less than clearance levels is radioactive from a physical viewpoint.

waste, transuranic Alpha-bearing waste containing nuclides with atomic numbers above 92, in quantities and/or concentrations above regulatory limits.

waste, vitrified *See* **waste glass.**

waste disposal *See* **disposal.**

waste disposal system Refers to the disposal environment as a whole, including the geological surroundings, the engineering system of a repository (e.g., barriers), and the waste packages.

waste form Waste in its physical and chemical forms after treatment.

waste generator The operating organization of a facility or an activity that produces waste.

waste glass The vitreous product that results from incorporating waste into a glass matrix. *See* also **glass.**

End Notes

[1] Among those groups are the Forum on Stakeholder Confidence, the Advisory Bodies to Government, the Regulators' Forum, the Working Party on Decommissioning and Dismantling, and the Integration Group for the Safety Case of Radioactive Waste Repositories.

[2] Nuclear Energy Agency, Radioactive Waste Management Committee, "Moving Forward with Geological Disposal of Radioactive Waste: An NEA RWMC Collective Statement," NEA-6433, OECD, Paris, 2008, pp. 7, 14. The other two quotations in this paragraph come from the same source.

[3] Most of this material consists of liquid and vitrified high-level radioactive waste (HLW) from reprocessing plants and spent nuclear fuel (SNF).

[4] The information contained in the tables found in the Overview section as well as in the detailed tables uses terminology provided by the in-country reviewers. Space limitations sometimes prevent expanding on the information provided.

[5] Although the regulatory responsibilities and arrangements vary from country to country, in all cases the regulator is an official governmental body. See Nuclear Energy Agency, "Regulating Long-Term Safety of Geologic Disposal," NEA-6182, OECD, Paris, 2007.

[6] U.S. Congress, Nuclear Waste Policy Act of 1982, Section 303. See also Advisory Panel on Alternative Means for Financing and Managing Radioactive Waste Facilities, "Managing Radioactive Waste—A Better Idea," December 1984.

[7] National Research Council, *The Disposal of Radioactive Waste on Land*, (Washington, D.C.: National Academy of Sciences Press, 1957).

[8] The United States abandoned its salt-centric siting strategy in 1976. Studies by the U.S. Geological Survey and the American Physical Society argued that what matters is the performance of the entire system of geological and engineered barriers. This view was adopted for the most part in 1979 by an interagency group created by President Jimmy Carter to develop an Administration-wide policy on managing radioactive waste. See Interagency Review Group on Nuclear Waste Management, *Report to the President*, (TID-29442), Washington D.C., 1979.

[9] One millisievert (mSv) equals 100 millirems.

[10] This definition is consistent with how the term is used by the International Commission on Radiological Protection and the regulators in Germany and the United Kingdom.

[11] In addition to satisfying regulatory requirements for protecting public health and safety, the implementing organization typically has to prepare environmental impact assessments and have them approved by relevant governmental authorities.

[12] In the detailed tables, the reader will find two entries that appear similar on their face: "no decision made" and "none." When the first entry is encountered, it should be interpreted to mean that the country has not addressed the particular attribute either implicitly or explicitly. When the second entry is encountered, it should be interpreted to mean that the country has made the decision indicated in the table.

CHAPTER SOURCES

The following chapters have been previously published:

Chapter 1 – This is an edited, reformatted and augmented version of Congressional Research Service publication, Report RL34234, dated March 5, 2010.

Chapter 2 – This is an edited, reformatted and augmented version of United States Government Accountability Office publication GAO-10-489, dated April 2010

Chapter 3 – This is an edited, reformatted and augmented version of United States Nuclear Waste Technical Review Board, dated October 2009.

INDEX

A

B

C

D

E

F

G

H

I

J

K

L

M

N

S

T

U

V

W

Y

Z